Minitab Guide for Moore's

The Basic Practice of Statistics

Third Edition

Betsy S. Greenberg

W.H. Freeman and Company
New York

MINITAB® is a registered trademark of Minitab, Inc. Output from MINITAB is printed with permission of Minitab, Inc., State College, PA.

Printed in the United States of America

ISBN: 0-7167-5887-3

First Printing 2003

Contents

Contents

Preface

This *Minitab Guide* accompanies *The Basic Practice of Statistics* (BPS) by David S. Moore. The manual is intended to be used with the statistical software package called Minitab. Minitab was originally developed in 1972 to help professors teach basic statistics. The software is now used in more than 2000 colleges and universities around the world. Minitab is an easy-to-use software that provides students with a wide range of basic and advanced data analysis capabilities.

In addition to its comprehensive statistical capabilities, Minitab offers high-resolution graphics that enable students to produce a comprehensive array of graphs. Minitab also offers powerful data management capabilities, allowing users to import data from other versions of Minitab, spreadsheets, databases and text files. Minitab relieves students of tedious statistical calculations and allows them to better understand statistical concepts.

Minitab is also the tool of choice for businesses of all sizes. It is used in 80 countries throughout the world; from start-ups to the Fortune 500 companies, including Ford Motor Company, 3M, AlliedSignal, General Motors and Lockheed Martin.

Minitab is available on a wide variety of computers, including mainframes and personal computers. This book is based on a pre-release version of Release 14 for Windows, the most recent version of Minitab available. If you are using a version different from Release 14, there may be slight differences in the menu interfaces. For further information about the software, contact

Minitab Inc.
3081 Enterprise Drive
State College, PA 16801 USA
Phone: (814) 238-3280
Fax: (814) 238-4383
e-mail: Info@minitab.com
URL: http://www.minitab.com

This *Minitab Guide* parallels the BPS. The *Minitab Manual* contains an introduction to Minitab plus a chapter corresponding to each chapter in BPS. In each chapter, we show how Minitab menu commands can be used to perform the statistical techniques described in BPS. In addition, each chapter includes exercises selected and modified from BPS that are appropriate to be done using Minitab. The numbering of the exercises refers to exercises in BPS. The Appendix describes session commands and provides a list by topic of Minitab session commands and menu equivalents that are referred to in this guide.

April, 2003

Introduction to Minitab

Topics to be covered in this chapter:

What Minitab Will Do for You
Different Versions of Minitab
Beginning and Ending a Minitab Session
The Minitab Worksheet
Minitab Commands
Minitab's Calculator
The Data Window and Entering Data
Opening, Saving, and Printing Files
Changing the Data
Getting Help

What Minitab Will Do for You

Before the widespread availability of powerful computers and prepackaged statistical software, tedious manual computations were routine in statistics courses. Today, computers have revolutionized data analysis, which is a fundamental task of statistics. Packages such as Minitab allow the computer to automate calculations and graphs. Minitab can perform a wide variety of tasks, from the construction of graphical and numerical summaries for a set of data to the more complicated statistical procedures and tests described in *The Basic Practice of Statistics*, by David S. Moore. Minitab will free you from mathematical calculations and allow you to concentrate more on the analysis of data and the interpretation of the results. In this supplement, we will refer to the textbook as BPS. The numbering of exercises refers to exercises in BPS.

Different Versions of Minitab

Minitab runs on Windows and Macintosh computers, as well as on most of the leading workstations and mainframe computers. This book is based on a prerelease version of Release 14 for Windows, the most recent version of Minitab available. Different versions may look slightly different on the screen and require different methods of executing commands. Windows and Macintosh versions have menus that allow you to choose commands. In addition, all versions of Minitab allow you to type commands. This book will illustrate menu commands in detail. Session commands are listed in the Appendix. If you are using a version other than Release 14, there may be some differences in the menu interfaces. Manuals that come with the software as well as online help are available to give you more information.

Beginning and Ending a Minitab Session

To start a Minitab session from the menu, select

Start ➤ Programs ➤ Minitab 14 for Windows ➤ Minitab

To exit Minitab, select

File ➤ Exit

from the menu. When you first enter Minitab, the screen will appear as in the figure with a toolbar, a Session window, and a Data window. Additional windows such as graph windows and dialog boxes may also appear as you use Minitab.

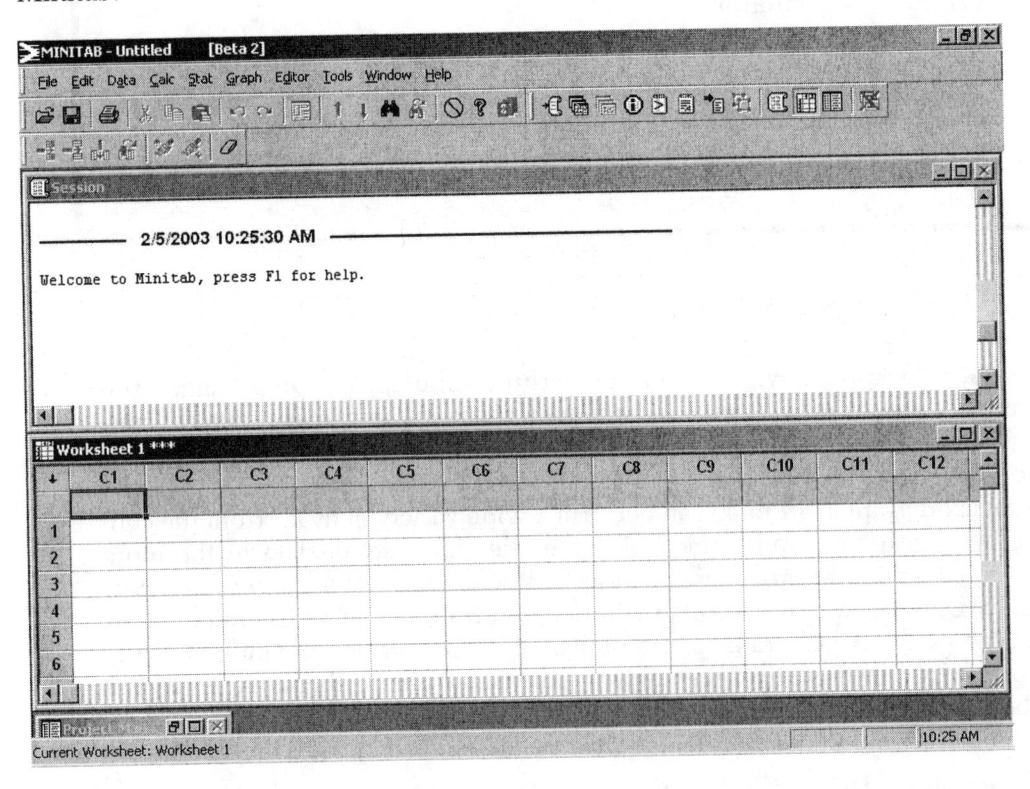

The Minitab Worksheet

The Minitab worksheet is arranged by rows and columns. The columns, C1, C2, C3, and so on, correspond to the variables in your data, the rows to observations. The columns can be viewed in the Data window. In addition, the worksheet may also include stored constants, K1, K2, K3, and so on. Most of the Minitab commands address the columns. In general, a column contains data for one variable, and each row contains all the data for a subject or observation. Columns can be referred to by number (C1, C2, C3, and so on.) or by name such as "height" or "weight."

In the Minitab worksheet, constants are referenced by the letter K and a number (K1, K2, K3, and so on). Unlike columns, constants are single values and do not show up in the data window. Storing a constant tells Minitab to remember this value; it will be needed later. Constants are analogous to the memory functions on most calculators. For instance, Minitab allows you to quickly find the average of a column of numbers. Instead of having to write it down, the value can be stored into a constant such as K1 and used in subsequent calculations.

Minitab Commands

Commands tell Minitab what to do. You can issue commands in Minitab by choosing commands from the menus or by typing session commands directly into the Session window. Menu commands are described throughout this guide. Session commands are an alternative to menu commands. They are described in the Appendix along with a list of session commands and their menu equivalents.

To issue a menu command, click on an item in the menu bar to open the menu, then click on a menu item to execute the command, open a submenu, or open a dialog box. For example, to use Minitab's calculator, select **Calc ➤ Calculator** from the menu. The following dialog box appears, prompting you for additional information needed to carry out the command.

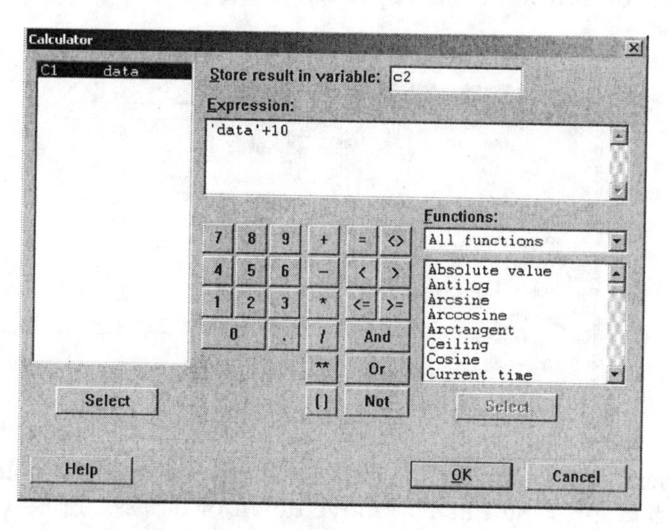

Dialog boxes may have text boxes, selection lists, and buttons. Many of the dialog boxes have buttons that lead to subdialog boxes. Most of Minitab's dialog boxes have a variable list in the upper left corner. The list will include columns and constants from the current worksheet. To enter a variable in a text box, click or place the cursor in a text box. The variable list will now display only the valid choices for the text box. Click on the desired variable in the variable box and then click select or simply double click on the desired variable. You can also type the desired column or variable name into the text box.

Minitab's Calculator

Minitab's calculator lets you perform mathematical operations and functions. The results of a calculation can be stored in a column or constant. To use the calculator, choose

Calc ➤ Calculator

from the menu. The dialog box shown on the previous page will appear. Under Store result in variable, enter a new or existing column or constant. Under Expression, select variables and functions from their respective lists, and click calculator buttons for numbers and arithmetic functions. You can also type the expressions.

Minitab's calculator performs the basic operations of addition (+), subtraction (–), multiplication (*), division (/), and exponentiation (**). Many additional Minitab functions such as can be used as part of the expression in the calculator's dialog box.

The Data Window and Entering Data

The Data window shows the columns in your worksheet and allows you to easily enter, edit, and view your data. To enter a value in a Data window cell, just click on the cell, type a value, and press Enter. To enter a column of data, click the data direction arrow in the upper left corner of the worksheet to make it point down.

Worksheet 1 ***

→	C1-T	C2	C3	C4	C5	C6
	Material	Weight				
1	Food Scraps	25.9				
2	Glass	12.8				
3	Metals	18.0				
4	Paper, paperboard	86.7				
5	Plastics					
6	Rubber, leather, textiles					
7	Wood					
8	Yard trimmings					
9	Other					
10						

To enter a row of data, click the data direction arrow to make it point to the right. Enter your data, pressing Enter to move down or across. Press Ctrl + Enter to move to the start of the next column or row.

You can copy from cells, rows, or columns of the same or another data window. You can also copy from other applications such as spreadsheets or word processors. To copy to the Data window select

Edit ➤ Copy Cells

from the Minitab menu or usually just **Edit ➤ Copy** for other applications. To paste onto the data window, click on the upper left cell of the area where you want the data pasted and then select

Edit ➤ Paste

from the Minitab menu.

Opening, Saving, and Printing Files

To open data from a file, select

File ➤ Open Worksheet

from the menu. In the Files of type box, choose the type of file you are looking for: Minitab, Minitab portable, Excel, and so on. Select a file, and click Open.

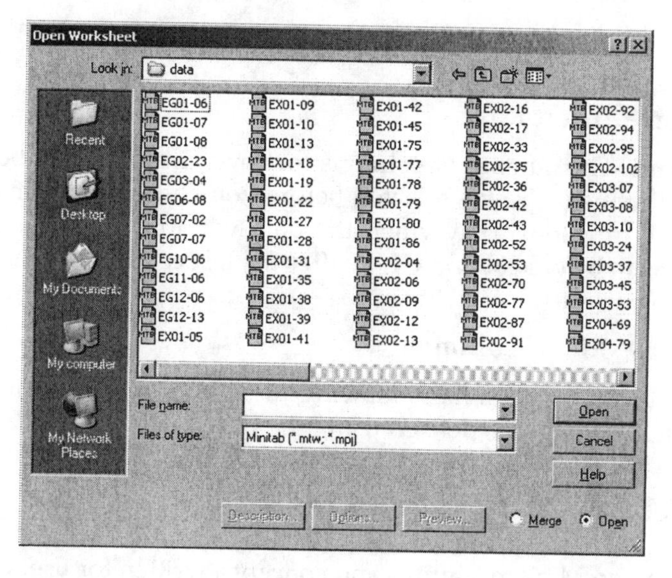

To print the contents of the data window, click on the data window. Then select

File ➤ Print Worksheet

from the menu Select the desired options in the dialog box and click OK.

Each data set you work with is contained in a worksheet. In Minitab Version 13, you can have many worksheets and graphs in one project. To save data as part of a project, select

File ➤ Save Project

from the menu. To save data into a separate file, make the desired Data window active, and select

File ➤ Save Worksheet As

from the menu. In Save as type, choose the data format in which you want the data to be saved. Select a directory, enter a file name, and click Save.

Changing the Data

Data in the data window can be corrected by simply clicking on a cell, typing in a correct entry, and hitting Enter. For more extensive changes, the Editor menu can be used. Under the Editor menu, you may choose to either insert cells, rows, or columns in the data set. To insert one or more empty cells above the active cell of the data window. select

Editor ➤ Insert Cells

from the menu. The remaining cells in the column move down. The number of cells inserted will be equal to the number of cells selected before you choose the command. To insert one or more empty columns to the left of the active column, select

Editor ➤ Insert Columns

from the menu. Similarly, to insert one or more empty rows above the active row, select

Editor ➤ Insert Rows

from the menu. The number of columns (or rows) inserted will be equal to the number of columns (or rows) selected before you choose the command. These commands are functional only when the data window is active. To move selected columns to before column C1, to after the last column in use, or to before a specific column select

Editor ➤ Move Columns

from the menu. To delete in the Data window, columns, rows, or individual cells can be highlighted and then deleted using the Delete key.

Getting Help

Documentation on Minitab features and concepts, written for users of menus and dialog boxes, can be obtained by clicking the Help button in any dialog box or pressing F1 at any time. For example, for help with Minitab's calculator, click on the Help button on the Calculator dialog box shown on page 3. Help can be obtained from the menu by selecting

Help ➤ Search Help

and then selecting a topic. A sample of the documentation available appears on the following page.

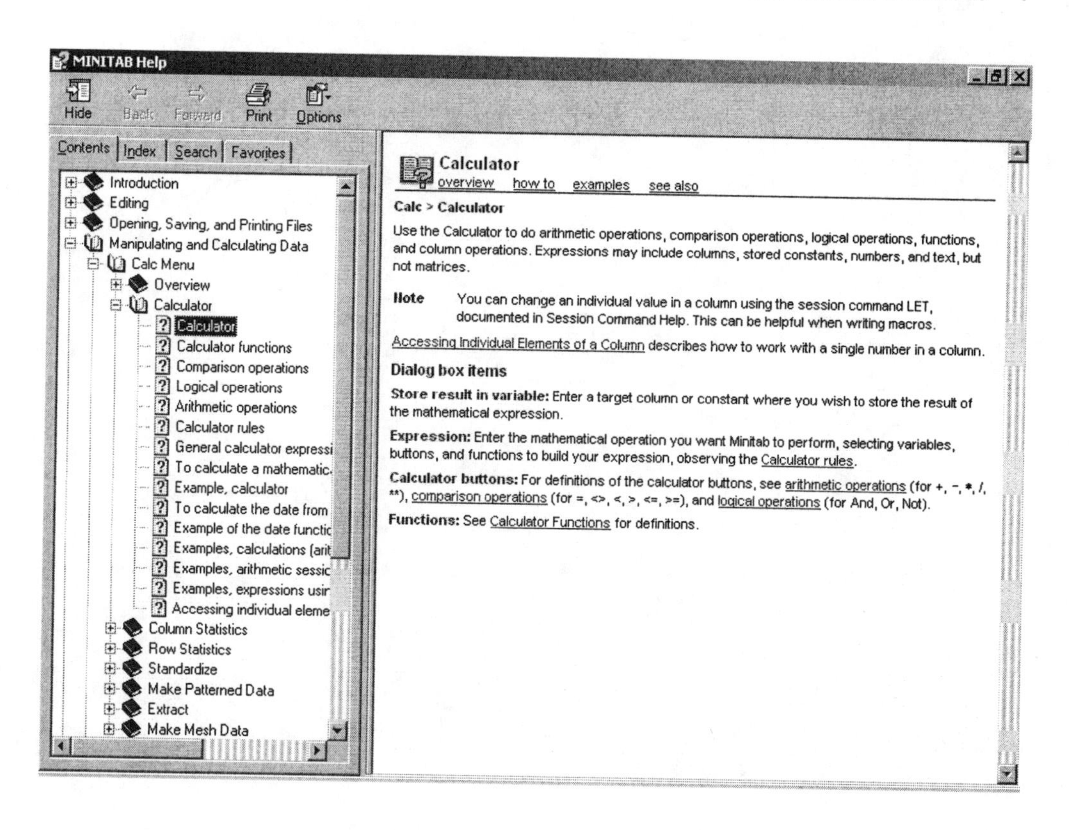
MINITAB Help
Hide
Back
Forward
Print
Options
Contents
Index
Search
Favorites
Introduction
Editing
Opening, Saving, and Printing Files
Manipulating and Calculating Data
Calc Menu
Overview
Calculator
Calculator
Calculator functions
Comparison operations
Logical operations
Arithmetic operations
Calculator rules
Column Statistics
Row Statistics
Standardize
Make Patterned Data
Extract
Make Mesh Data
Calculator
overview how to examples see also
Calc > Calculator
Use the Calculator to do arithmetic operations, comparison operations, logical operations, functions, and column operations. Expressions may include columns, stored constants, numbers, and text, but not matrices.
Note You can change an individual value in a column using the session command LET, documented in Session Command Help. This can be helpful when writing macros.
Accessing Individual Elements of a Column describes how to work with a single number in a column.
Dialog box items
Store result in variable: Enter a target column or constant where you wish to store the result of the mathematical expression.
Expression: Enter the mathematical operation you want Minitab to perform, selecting variables, buttons, and functions to build your expression, observing the Calculator rules.
Calculator buttons: For definitions of the calculator buttons, see arithmetic operations (for +, −, *, /, **), comparison operations (for =, <>, <, >, <=, >=), and logical operations (for And, Or, Not).
Functions: See Calculator Functions for definitions.

Chapter 1
Picturing Distributions with Graphs

Topics to be covered in this chapter:

Bar Charts
Pie Charts
Histograms
Stemplots
Time Series Plots

Bar Charts

Minitab allows us to examine the distribution of variables with graphs. Bar charts are useful for categorical data. We will use the data found in Example 1.2 in BPS and EG01-02.MTW to show how to make a bar chart with Minitab. The data give the breakdown of the materials that made up American municipal solid waste in 2000. The materials are entered in C1 and the weights (in millions of tons) are given in C2. To make a bar chart of the data, select

Graph ➤ Bar Chart

from the menu. Select Values from a table and Simple and then click OK.

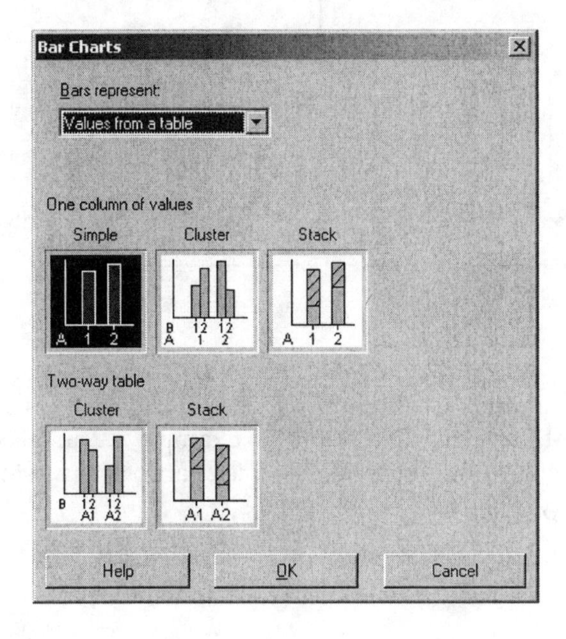

As shown in the following dialog box, the graph variable and categorical variables must be filled in. In this example, the categorical variable is "Material" and the graph variable is "Weight".

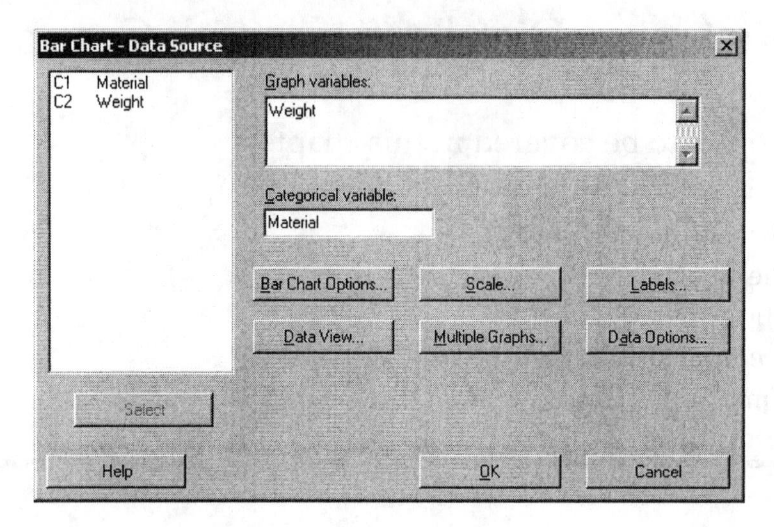

Clicking on OK will produce the following bar chart.

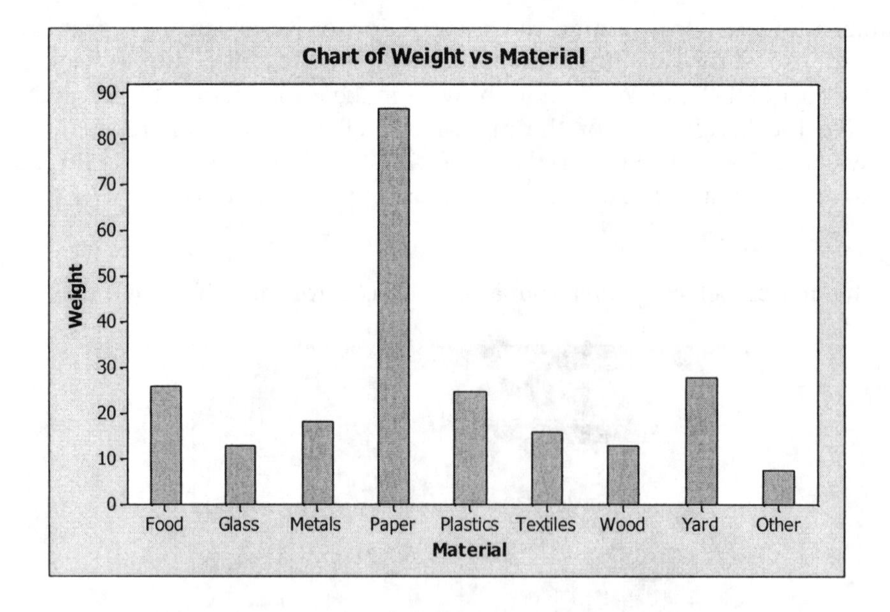

Pie Charts

Another way to examine distributions of categorical variables is with a pie chart. We will continue to use the data found in Example 1.2 in BPS to show how to make a pie chart with Minitab. To make a pie chart of the waste data, select

Graph ➤ Pie

from the menu. Because the data are tabulated, click on Chart values from a table on the dialog box and fill in the appropriate variables for the categorical and summary variables.

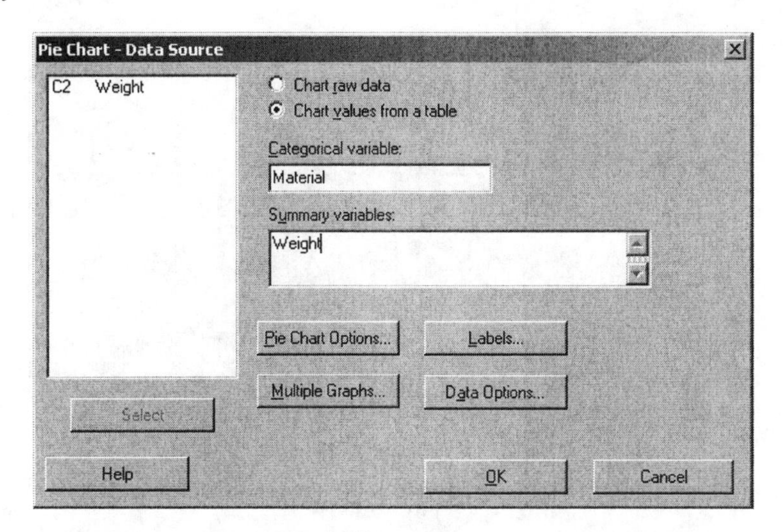

The pie chart for Example 1.2 follows.

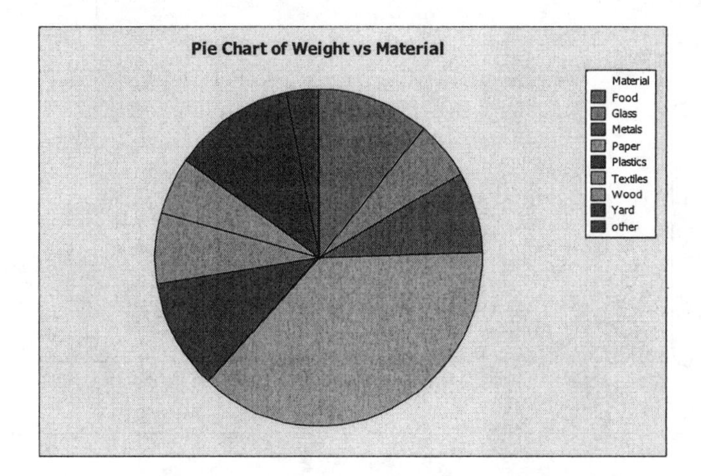

Histograms

The most common graph for the distribution of a quantitative variable is a histogram. Example 1.8 in BPS and EG01-08.MTW give data from a typical student laboratory exercise: the load in pounds needed to pull apart pieces of Douglas fir 4 inches long and 1.5 inches square. To create a histogram for these data, select

Graph ➤ Histogram

from the menu. Click on Simple and then OK in the first dialog box. The next dialog box will appear as follows. In the dialog box double click on the variable named load and click on OK.

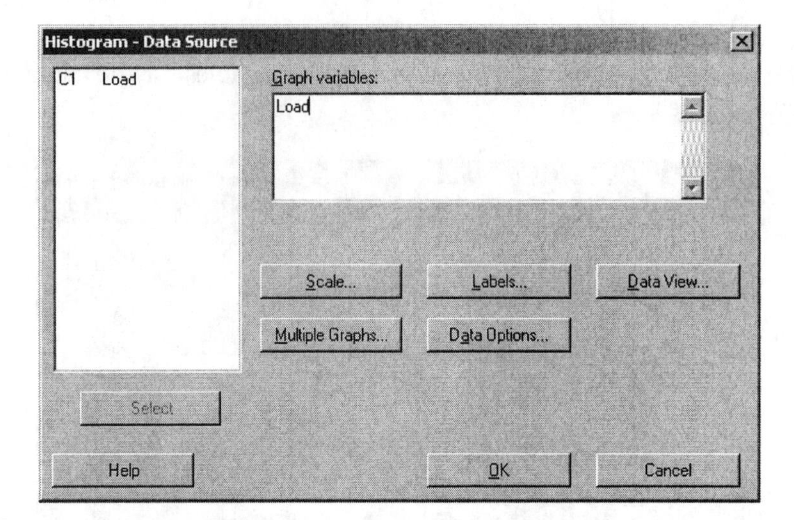

The histogram appears as follows.

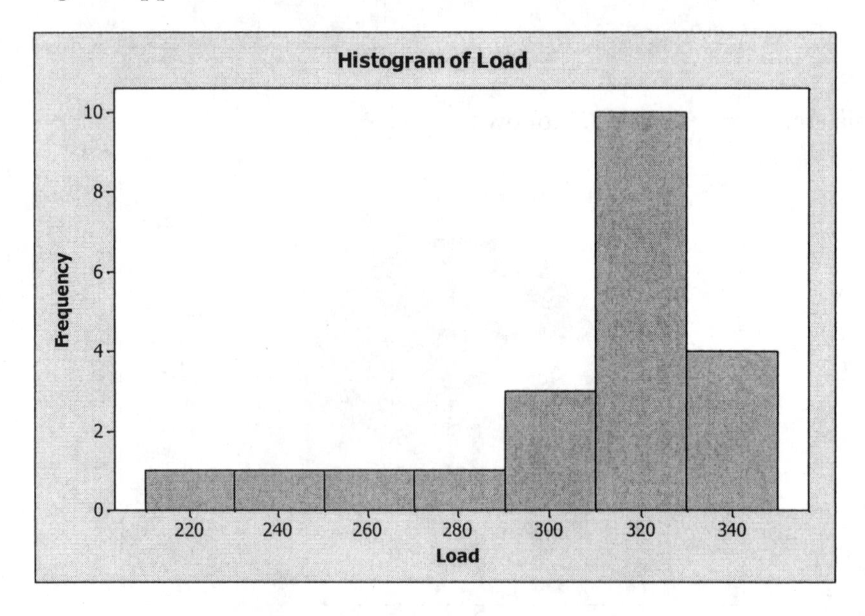

To change the type of histogram from frequency to percent, click on the Scales button and then the Y-Scale Type tab and click on Percent.

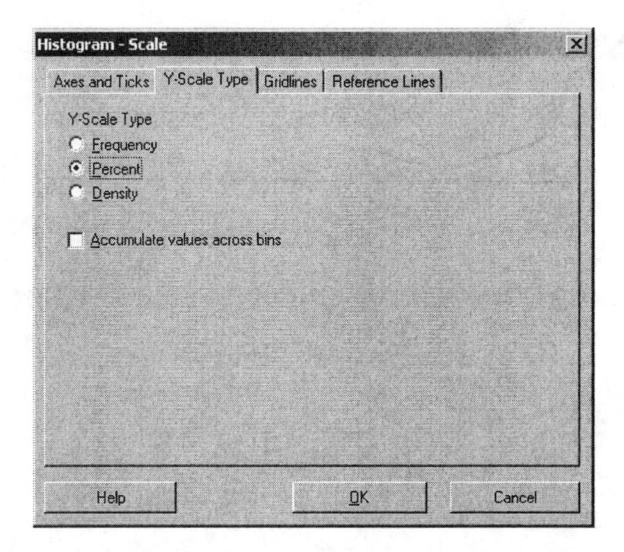

You may also change from Minitab's automatic choice of intervals. Once you've produced a histogram, double click on the *X*-scale to obtain the Edit Scale dialog box. Click on the Binning tab to change from midpoints to cutpoints and/or specify midpoint/cutpoint positions.

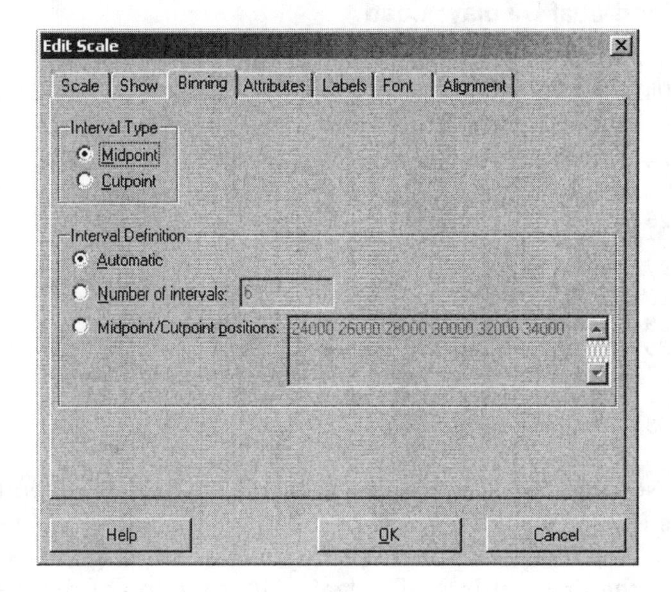

Stemplots

A stem-and-leaf plot (or stemplot) uses the actual data from a quantitative distribution to create the display. To create the stemplot, select

Graph ➤ Stem-and-Leaf

from the menu. The session window will appear as follows.

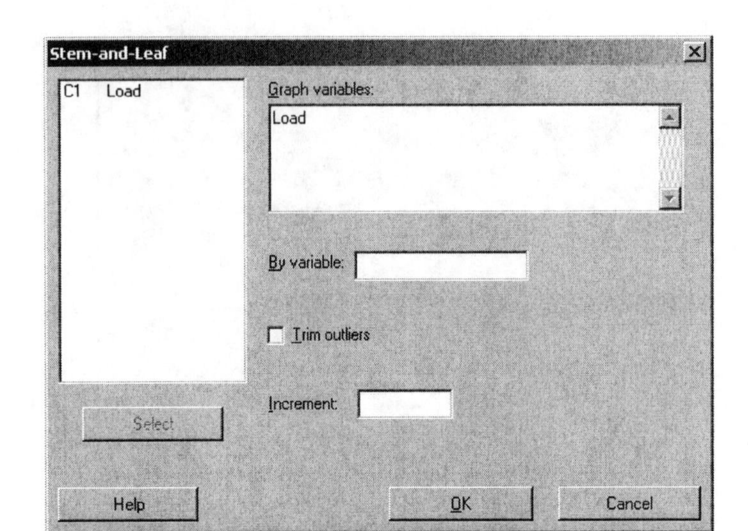

When you click on the Graph Variables text box, the valid choices appear in the variable list box. Click on the desired variable and then click Select. When you click OK, the following stemplot will appear in the Sesson window.

Stem-and-Leaf Display: Load

```
Stem-and-leaf of Load  N  = 20
Leaf Unit = 100


 1    23  0
 2    24  0
 2    25
 3    26  5
 3    27
 4    28  7
 4    29
 7    30  149
10    31  389
10    32  033577
 4    33  0126
```

The first column of a stem-and-leaf display is called the depth, the second column holds the stems, and the rest of the display holds the leaves. Each leaf digit represents one observation. In the stem-and-leaf display shown, the first stem is 23 and the first leaf is 0. The leaf unit at the top of the display tells us where to put the decimal point. In this example, the Leaf Unit = 100. Therefore the corresponding first observation is 23,000 (which was rounded from 23,040.)

The column on the left gives a cumulative count of values from the top of the figure down and from the bottom of the figure up to the middle. The count for the row containing the median has parentheses around it. Parentheses around the median row are omitted if the median falls between two lines of the display. This occurred in the stemplot shown.

If you wish, you can control the scaling of a stem-and-leaf display by specifying an increment. For example, we can choose an increment of 20 to obtain the following display.

Stem-and-Leaf Display: Load

```
Stem-and-leaf of Load  N  = 21
Leaf Unit = 10


 1    2  3
 2    2  4
 3    2  6
 4    2  8
 10   3  000111
(11)  3  22222223333
```

Time Series Plots

When quantitative data are collected over time, it is a good idea to plot the observations in the order they were collected. To create a time plot, select

Graph ➤ Time Series Plot

from the menu. This command plots time series data on the y-axis versus time on the x-axis. For example, the data in Table 1.3 in BPS and TA01-03.MTW give the average tuition and fees paid by college students at four-year colleges, both public and private, from the 1971–1972 academic year to the 2001–2002 academic year.

To make a time series plot of this data, select **Graph ➤ Time Series Plot** from the menu. Since there are two time series to plot, choose Multiple and then click on OK in the first dialog box. The next dialog box will appear as follows.

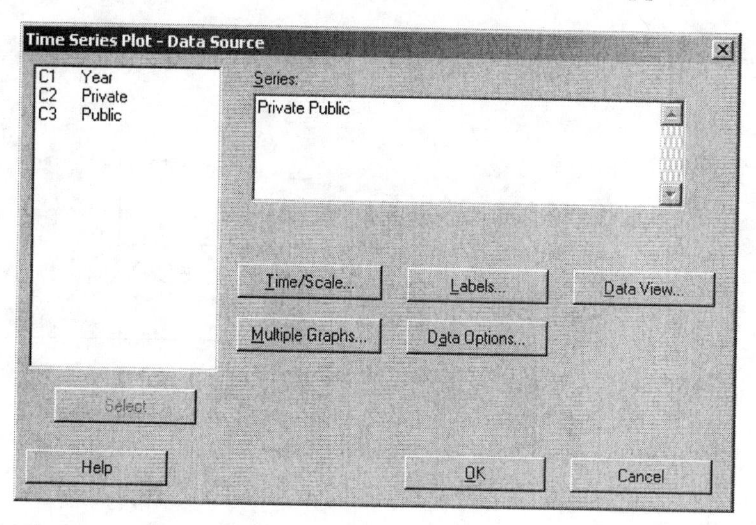

Select the series (Private and Public) to be plotted and then click on the Time/Scale button. In the Time/Scale subdialog box, we specify that the Calendar is based on Year, starting in 1971 with an increment of 1.

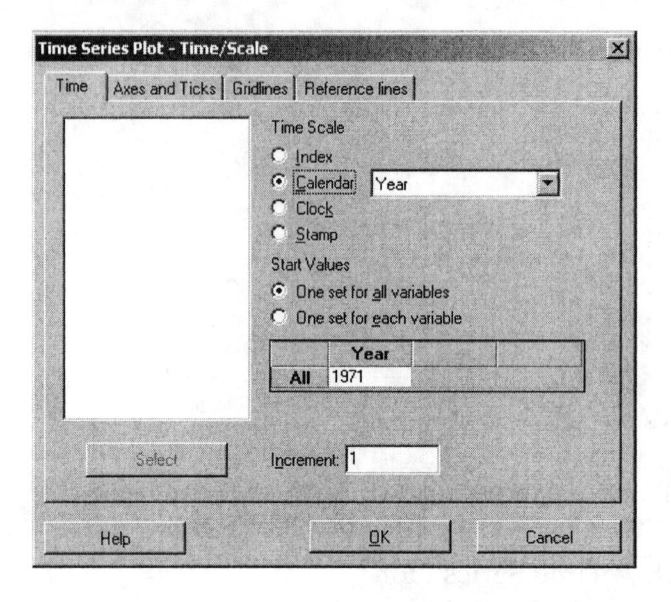

Click on OK in the subdialog box and then in the dialog box to obtain the following time series plot.

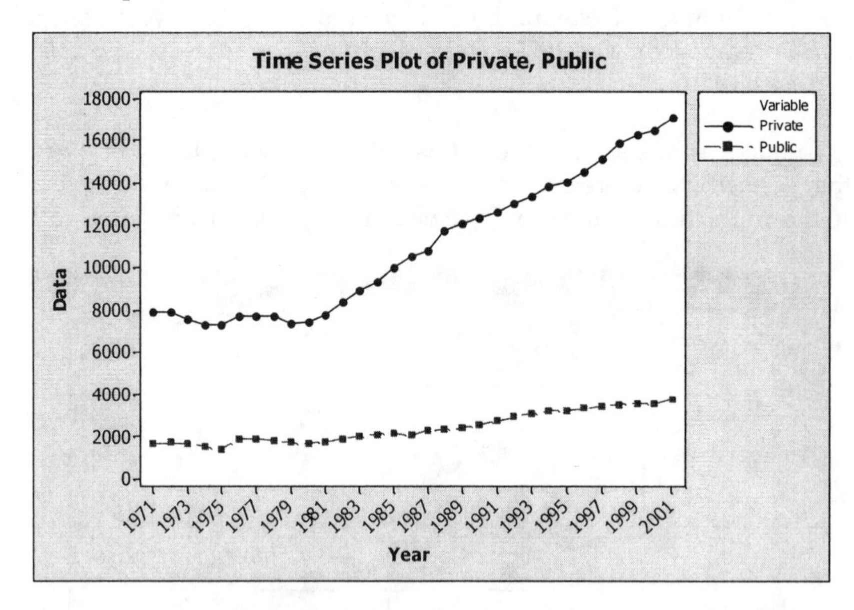

EXERCISES

1.3 The color of your car. Here is a breakdown of the most popular colors for vehicles made in North America during the 2001 model year.

Color	Percent
Silver	21.0%
White	15.6%
Black	11.2%
Blue	9.9%
Green	7.6%
Medium red	6.9%
Brown	5.6%
Gold	4.5%
Bright red	4.3%
Grey	2.0%

(a) Select **Graph ➤ Bar Chart** to make a bar chart of the color data.

(b) What percent of vehicles have some other color? Select **Calc ➤ Column Statistics** and then click on Sum to help you find out. Add an "Other" category to your worksheet. Check that the sum in the Percent column is now 100.

(c) Select **Graph ➤ Pie Chart** from the menu to make a pie chart of the color data.

1.4 Births are not, as you might think, evenly distributed across the days of the week. Here are the average numbers of babies born on each day of the week in 1999.

Day	Births
Sunday	7,731
Monday	11,018
Tuesday	12,424
Wednesday	12,183
Thursday	11,893
Friday	12,012
Saturday	8,654

(a) Select **Graph ➤ Bar Chart** to present these data in a well-labeled bar chart.

(b) Select **Graph ➤ Pie Chart** to make a pie chart of these data. Suggest some possible reasons why there are fewer births on weekends.

1.6 Table 1.2 in BPS and TA01-02.MTW give data on the fuel economy of 1998 model midsize cars.

(a) Select **Graph ➤ Histogram** to make a histogram of the data. Your histogram should show an extreme high outlier. This is the

Honda Insight, a hybrid gas-electric car that is quite different from the others listed.

(b) Make a new histogram of highway mileage, leaving out the Insight. Describe the main features (shape, center, spread, outliers) of the distribution of highway mileage.

(c) The government imposes a "gas guzzler" tax on cars with low gas mileage. Which of these cars do you think may be subject to the gas guzzler tax?

1.8 The Survey of Study Habits and Attitudes (SSHA) is a psychological test that evaluates college students' motivation, study habits, and attitudes toward school. A private college gives the SSHA to 18 of its incoming first-year women students. Their scores are

154	109	137	115	152	140	154	178	101
103	126	126	137	165	165	129	200	148

Enter the data into a Minitab worksheet and select **Graph ➤ Stem-and-Leaf** from the menu to make a stemplot of these data. The overall shape of the distribution is irregular, as often happens when only a few observations are available. Are there any outliers? About where is the center of the distribution (the score with half the scores above it and half below)? What is the spread of the scores (ignoring any outliers)?

1.9 Select **Graph ➤ Stem-and-Leaf** from the menu to make stemplots of the data on Hispanics found in Table 1.1 and TA01-01.MTW. Make three different stemplots by letting the increment be equal to 1, 5, and then 10. Which stemplot do you prefer? Why?

1.10 Garbage that is not recycled is buried in landfills. Here and in EX01-10.MTW are time series data that emphasize the need for recycling: the number of landfills operating in the United States in the years 1988 to 2000.

Year	Landfills	Year	Landfills	Year	Landfills
1988	7924	1993	4482	1997	2514
1989	7379	1994	3558	1998	2314
1990	6326	1995	3197	1999	2216
1991	5812	1996	3091	2000	1967
1992	5386				

Select **Graph ➤ Time Series Plot** from the menu to make a time plot of these data. Describe the trend that your plot shows. Why does the trend emphasize the need for recycling?

1.11 Car colors in Japan. Exercise 1.3 gives data on the most popular colors for motor vehicles made in North America. Here are similar data for 2001 model year vehicles made in Japan.

Color	Percent

Gray	43%
White	35%
Black	8%
Blue	7%
Red	4%
Green	2%

Select **Graph ➤ Histogram** to make a graph of these data. What are the most important differences between choice for vehicle colors in Japan and North America?

1.12 Deaths among young people. The number of deaths among persons aged 15 to 24 years in the United States in 2000 due to the leading causes of death for this age group were: accidents, 13,616; homicide, 4796; suicide, 3877; cancer, 1668; heart disease, 931; congenital defects, 425.

(a) Select **Graph ➤ Bar Chart** to display these data.

(b) What additional information do you need to make a pie chart?

1.19 Marijuana and traffic accidents. Researchers in New Zealand interviewed 907 drivers at age 21. They had data on traffic accidents and they asked their subjects about marijuana use. Here are data on the numbers of accidents caused by these drivers at age 19 broken down by marijuana use at the same age.

	Marijuana use per year			
	Never	1–10 times	11–50 times	50+ times
Drivers	452	229	70	156
Accidents caused	59	36	15	50

(a) Explain carefully why a useful graph must compare rates (accidents per driver) rather than counts of accidents in the four marijuana use classes.

(b) Enter the data into a Minitab worksheet. Let C1 be "use/year", C2 be "drivers", and C3 be "accidents caused". Select **Calc ➤ Calculator**, to find the number of accidents caused per driver and save the results in C4. Let C4 be "accident rates". Select **Graph ➤ Bar Chart** to make a graph that displays the accident rates for each class.

(c) What do you conclude? (You can't conclude that marijuana use causes accidents, because risktakers are more likely both to drive aggressively and to use marijuana.)

1.23 Table 1.4 in BPS and TA01-04.MTW give the percents of people living below the poverty line in the 26 states east of the Mississippi River. Select **Graph ➤ Stem-and-Leaf** to make a stemplot of these data. Is the distribution roughly symmetric, skewed to the right, or skewed to the left? Which states (if any) are outliers?

1.25 Here are the numbers of home runs that Babe Ruth hit in his 15 years with the New York Yankees, 1920 to 1934:

54 59 35 41 46 25 47 60 54 46 49 46 41 34 22

Enter the data into a Minitab worksheet. Select **Graph ➤ Stem-and-Leaf** from the menu to make a stemplot for these data. Is the distribution roughly symmetric, clearly skewed, or neither? About how many home runs did Ruth hit in a typical year? Is his famous 60 home runs in 1927 an outlier?

1.26 The leading contemporary home run hitter is Mark McGwire, who retired after the 2001 season. Here are McGwire's home run counts for 1987 to 2001:

49 32 33 39 22 42 9 9 39 52 58 70 65 32 29

Enter the data into a Minitab worksheet. Select **Graph ➤ Stem-and-Leaf** from the menu to make a stemplot for these data. If you didn't already solve problem 1.25, do the same for the Babe Ruth data given in exercise 1.25. Write a brief comparison of Ruth and McGwire as home run hitters. McGwire was injured in 1993 and there was a baseball strike in 1994. How do these events appear in the data?

1.27 Women were allowed to enter the Boston Marathon in 1972. The times (in minutes, rounded to the nearest minute) for the winning woman from 1972 to 2002 appear in Table 1.5 and TA01-05.MTW. In 2002, Margaret Okayo of Kenya set a new women's record for the race, 2 hours, 20 minutes, and 43 seconds.

(a) Select **Graph ➤ Time Series Plot** to make a time plot of the winning times.

(b) Give a brief description of the pattern of Boston Marathon winning times over these years. Has the rate of improvement in times slowed in recent years?

1.29 Table 1.6 and TA01-06.MTW give the number of medical doctors per 100,000 people in each state.

(a) Why is the number of doctors per 100,000 people a better measure of the availability of health care than a simple count of the number of doctors in a state?

(b) Make a graph that displays the distribution of doctors per 100,000 people. Write a brief description of the distribution. Are there any outliers? If so, can you explain them?

Chapter 2
Describing Distributions with Numbers

Topics to be covered in this chapter:

Numerical Descriptions
Boxplots

Numerical Descriptions

Numerical measures are often used to describe distributions. Select

Stat ➤ Basic Statistics ➤ Display Descriptive Statistics

from the menu to obtain descriptive statistics. The command summarizes several different measures of both the center and spread of a distribution. The command prints the statistics N, N*, Mean, SE Mean, StDev, Minimum, Q1, Median, Q3, and Maximum for each column specified.

If we want descriptive information for the 15 college graduates chosen at random by the Census Bureau in March 2002 and asked how much they earned in 2001, enter the following data into a Minitab worksheet or open EG02_01.MTW.

110 25 50 50 55 30 35 30 4 32 50 30 32 74 60

Descriptive information for the college graduates is obtained by selecting **Stat ➤ Basic Statistics ➤ Display Descriptive Statistics** from the menu. We select C1 or earnings for our variable in the dialog box as follows.

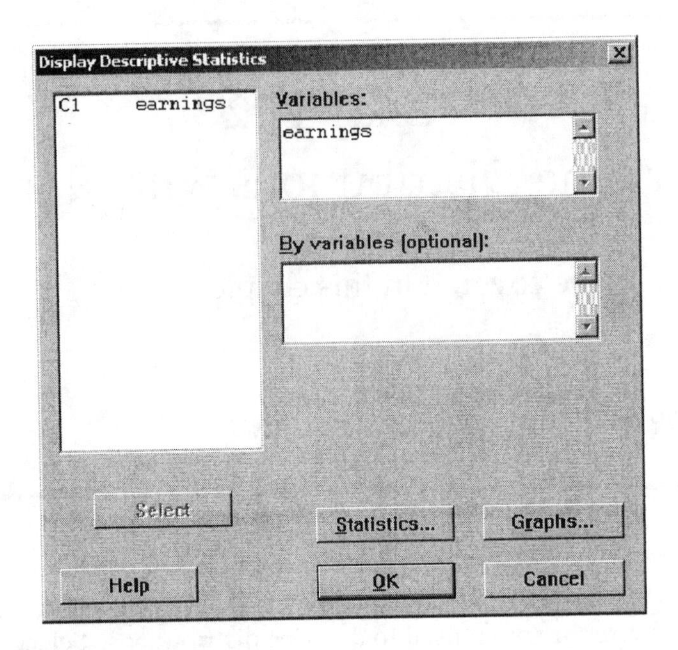

Descriptive Statistics: earnings

Variable	N	N*	Mean	SE Mean	StDev	Minimum	Q1	Median	Q3	Maximum
earnings	15	0	44.47	6.42	24.86	4.00	30.00	35.00	55.00	110.00

N is the number of actual values in the column (missing values are not counted). N* is the number (if any) of missing values. Mean is the average of the values. To find the median, the data first must be ordered. If N is odd, the median is the value in the middle. If N is even, the median is the average of the two middle values. StDev is the standard deviation computed as

$$\text{StDev} = \sqrt{\frac{\sum (x - \bar{x})}{\text{N} - 1}}$$

SE Mean is the standard error of the mean. It is calculated as $\text{StDev}/\sqrt{\text{N}}$. Q3 is the third quartile and Q1 is the first quartile. Minitab doesn't use exactly the same algorithm to calculate quartiles as BPS, so minor differences in results will sometimes occur.

The By variables box can be filled in to display descriptive statistics separately for each value of the specified variable. Column lengths must be equal to use the By variables box.

Column statistics can also be obtained individually by selecting

Calc ➤ Column Statistics

from the menu. The variable to be described and the descriptive measure or measures are selected on the dialog box.

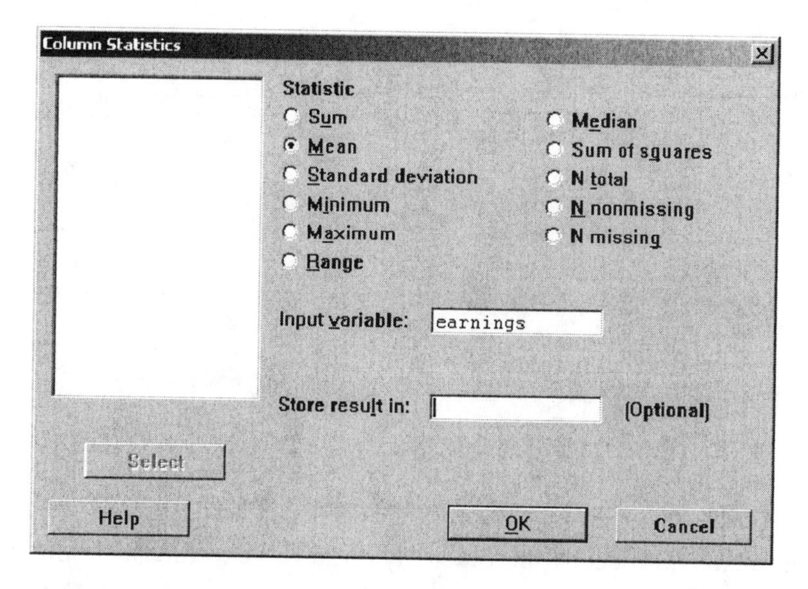

The statistics just described are also available for rows. These commands compute summaries across rows rather than down columns. The answers are always stored in a column. Rowwise statistics are obtained by selecting

Calc ➤ Row Statistics

from the menu. For the rowwise statistics as well as the column statistics, missing observations are omitted from the calculations.

Boxplots

The five-number summary consisting of the median, quartiles, and minimum and maximum values provides a quick overall description of a distribution. Boxplots based on the five-number summary display the main features of a column of data. Boxplots can be obtained by selecting

Graph ➤ Boxplot

from the menu.

A boxplot graphically displays the main features of data from a single variable. A boxplot is illustrated on the following page on the hourly bank workers. The boxplot consists of a box, whiskers, and outliers. Minitab draws a line across the box at the median. By default, the bottom of the box is at the first quartile (Q1) and the top is at the third quartile (Q3). The whiskers are the lines that extend from the top and bottom of the box to the adjacent values, the lowest and highest observations still inside the region defined by the lower limit $Q1-1.5(Q3-Q1)$ and the upper limit $Q3+1.5(Q3-Q1)$. Outliers are points outside the lower and upper limits, plotted with asterisks (*).

Selecting **Graph ➤ Boxplot** from the menu and filling in the dialog box will produce a boxplot(s). For a boxplot of one variable, select a column for the box below Y. This is illustrated for the earnings data from Example 2.1 in BPS and EG02-01.MTW.

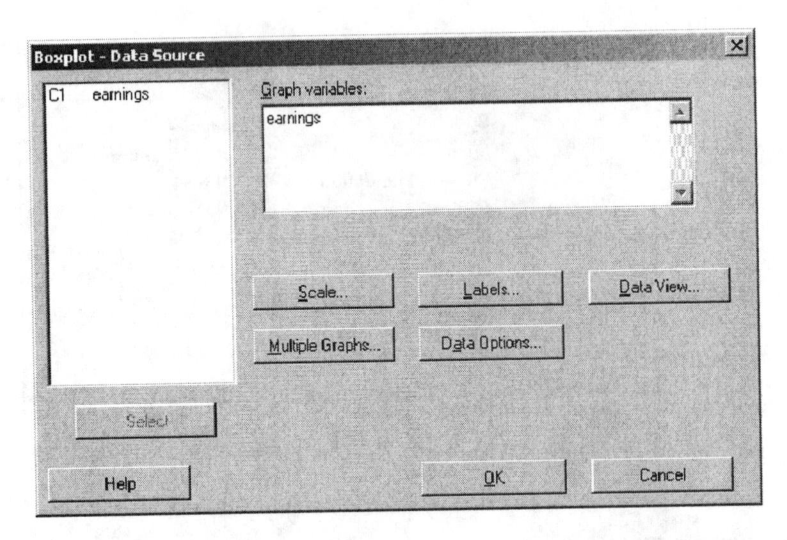

In the boxplot below an unusually large value (outlier) is plotted with an asterisk (*).

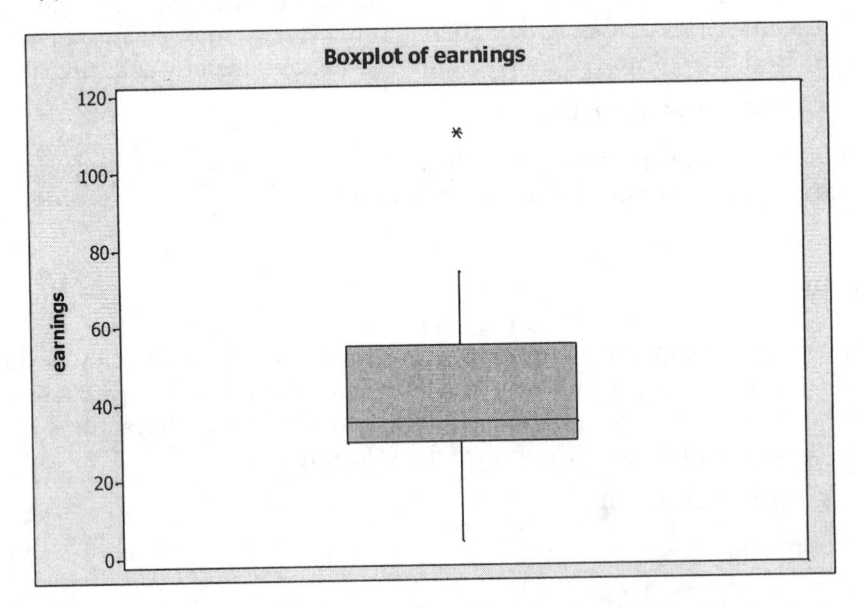

Example 2.5 in BPS and EG02-05.MTW give earnings data for a sample of high school graduates. To construct side-by-side boxplots comparing the earnings of the college graduates with the earnings of the high school graduates copy the high school graduates' data onto EG02-01.MTW. The data can be entered into a separate column. Select **Graph ➤ Boxplot** from the menu and click on Multiple Y's Simple in the first dialog box. The next dialog box will then give space for the graph variables as follows. We select the columns with the college and high school data and click OK to obtain side-by-side boxplots.

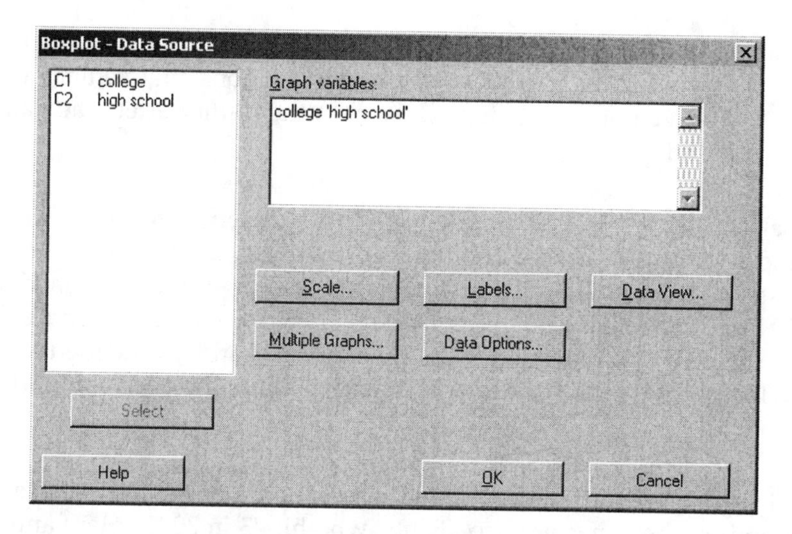

Side-by-side boxplots comparing the earnings of the college graduates with the earnings of the high school graduates appear below.

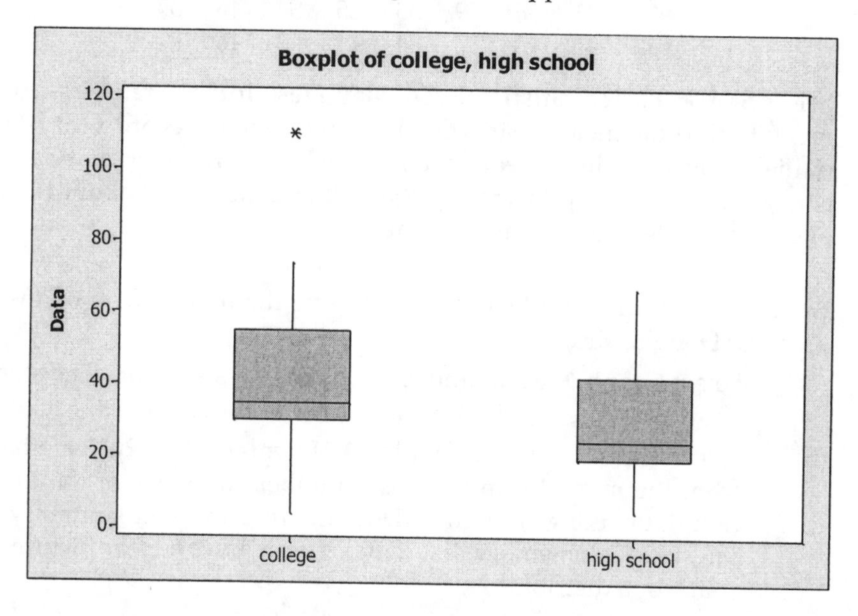

As an alternative, data for side-by-side boxplots can be arranged with the measurements (earnings) in one column and a categorical variable (education level) in another. In this case, select **Graph ➤ Boxplots** from the menu and then One Y With Groups in the first dialog box to make side-by-side boxplots.

EXERCISES

2.1 Table 1.2 and TA01-02.MTW give the gas mileages for the 22 two-seater cars listed in the government's fuel economy guide.

(a) Select **Stat ➤ Basic Statistics ➤ Display Descriptive Statistics** from the menu to find the mean $\bar{x}$ and the standard deviation s.

(b) The Honda Insight is an outlier that doesn't belong with the other cars. Find $\overline{x}$ and s for the observations that remain when you leave out the outlier. How does the outlier affect the values of $\overline{x}$ and s?

2.2 Select **Stat ➤ Basic Statistics ➤ Display Descriptive Statistics** from the menu to find the median highway mileage for the 22 two–seater cars listed in Table 1.2 of BPS and TA01-02.MTW. What is the median of the 21 cars that remain if we remove the Honda Insight? Compare the effect of the Insight on mean mileage (Exercise 2.1) and on the median mileage. What general fact about the mean and median does this comparison illustrate?

2.4 The major league baseball single-season home run record is held by Barry Bonds of the San Francisco Giants, who hit 73 in 2001. Here and in EX02-04.MTW are Bonds's home run totals from 1986 (his first year) to 2002:

16 25 24 19 33 25 34 46 37
33 42 40 37 34 49 73 46

Select **Stat ➤ Basic Statistics ➤ Display Descriptive Statistics** from the menu to find the mean $\overline{x}$ and the median. Bonds's record year is a high outlier. How do his career mean and median number of home runs change when we drop the record 73? What general fact about the mean and median does your result illustrate?

2.5 Example 1.8 in BPS and EG01-08.MTW give the breaking strengths of 20 pieces of Douglas fir.

(a) Select **Graph ➤ Stem-and-Leaf** to make a stemplot. The stemplot shows that the distribution is skewed to the left.

(b) Select **Stat ➤ Basic Statistics ➤ Display Descriptive Statistics** from the menu to find the five-number summary of the distribution of breaking strengths. Does the five–number summary show the skew? Remember that only a graph gives a clear picture of the shape of a distribution.

2.6 Table 2.1 in BPS and TA02-01.MTW give the city and highway gas mileage for 36 midsize cars from the 2002 model year. There is one low outlier, the 12-cylinder Rolls-Royce. We wonder if midsize sedans get better mileage than sports cars.

(a) Select **Stat ➤ Basic Statistics ➤ Display Descriptive Statistics** from the menu to find the five-number summaries for both city and highway mileage for the midsize cars in TA02-01.MTW and for the two-seater cars in TA01-02.MTW. (Leave out the Honda Insight.)

(b) Copy the data from TA01-02.MTW onto TA02-01.MTW. Select **Graph ➤ Boxplot** to make four side-by-side boxplots to display the summaries. Write a brief description of city versus highway and two-seaters versus midsize cars.

2.7 How old are presidents at their inauguration? Was Bill Clinton, at age 46, unusually young? Table 2.2 in BPS and TA02-02.MTW give the data, the ages of all U.S. presidents when they took offce.

(a) Select **Graph ➤ Stem-and-Leaf** to make a stemplot of the distribution of ages. From the shape of the distribution, do you expect the median to be much less than the mean, about the same as the mean, or much greater than the mean?

(b) Select **Stat ➤ Basics Statistics ➤ Desplay Descriptive Statistics** to find the mean and the five-number summary. Verify your expectation about the median.

(c) What is the range of the middle half of the ages of new presidents? Was Bill Clinton in the youngest 25%?

2.9 The mean $\bar{x}$ and standard deviation s measure center and spread but are not a complete description of a distribution. Data sets with different shapes can have the same mean and standard deviation. To demonstrate this fact, select **➤ Basics Statistics ➤ Desplay Descriptive Statistics** to find $\bar{x}$ and s the two small data sets found below and in EX02-09.MTW. Then select **Graph ➤ Stem-and-Leaf** to make a stemplot of each and comment on the shape of each distribution.

Data A	9.14	8.14	8.74	8.77	9.26	8.10	6.13	3.10	9.13	7.26	4.74
Data B	6.58	5.76	7.71	8.84	8.47	7.04	5.25	5.56	7.91	6.89	12.50

2.12 How much do users pay for Internet access? Here and in EX02-12.MTW are the monthly fees (in dollars) paid by a random sample of 50 users of commercial Internet service providers in August 2000:

20	40	22	22	21	21	20	10	20	20
20	13	18	50	20	18	15	8	22	25
22	10	20	22	22	21	15	23	30	12
9	20	40	22	29	19	15	20	20	20
20	15	19	21	14	22	21	35	20	22

(a) Select **Graph ➤ Stem-and-Leaf** to make a stemplot of these data. Briefly describe the pattern you see. About how much do you think America Online and its larger competitors were charging in August 2000? Are there any outliers?

(b) To report a quick summary of how much people pay for Internet service, do you prefer $\bar{x}$ and s or the five-number summary? Why? Select **Stat ➤ Basic Statistics ➤ Display Descriptive Statistics** to calculate your preferred summary.

2.13 Table 1.6 in BPS and TA01-06.MTW give the number of medical doctors per 100,000 people in each state. Exercise 1.29 asked you to plot the data. The distribution is right-skewed with several high outliers.

(a) Do you expect the mean to be greater than the median, about equal to the median, or less than the median? Why? Select **Stat ➤ Basic Statistics ➤ Display Descriptive Statistics** to calculate $\bar{x}$ and M and verify your expectation.

(b) The District of Columbia is a high outlier at 758 M.D.'s per 100,000 residents. If you remove D.C. because it is a city rather than a state, do you expect $\bar{x}$ or M to change more? Why? Select **Stat ➤ Basic Statistics ➤ Display Descriptive Statistics** to calculate both measures for the 50 states (omitting D.C.) and verify your expectation.

2.14 Exercises 1.25 and 1.26 give the numbers of home runs hit each season by Babe Ruth and Mark McGwire. Enter the data for each player into a separate column on the same Minitab worksheet. Select **Stat ➤ Basic Statistics ➤ Display Descriptive Statistics** to find the five-number summaries for each hitter. Select **Graph ➤ Boxplot** and make side-by-side box plots to compare these two home run hitters. What do your plots show?

2.15 Here are the scores of 18 first-year college women on the Survey of Study Habits and Attitudes (SSHA):

154	109	137	115	152	140	154	178	101
103	126	126	137	165	165	129	200	148

(a) Enter the data into a Mintab worksheet and select **Calc ➤ Column Statistics** to obtain the mean.

(b) A stemplot (Exercise 1.8) suggests that the score 200 is an outlier. Select **Calc ➤ Column Statistics** to find the mean for the 17 observations that remain when you drop the outlier. How does the outlier change the mean?

2.17 In Exercise 2.15 you found the mean of the SSHA scores of 18 first-year college women. Now select **Calc ➤ Column Statistics** to find the median of these scores. Is the median smaller or larger than the mean? Explain why this is so.

2.19 Does breast-feeding weaken bones? Breast-feeding mothers secrete calcium into their milk. Some of the calcium may come from their bones, so mothers may lose bone mineral. Researchers compared 47 breast-feeding women with 22 women of similar age who were neither pregnant nor lactating. They measured the percent change in the mineral content of the women's spines over three months. Here and in EX02-19.MTW are the data:

Breast-feeding women						Other women					
−4.7	−2.5	−4.9	−2.7	−0.8	−5.3	2.4	0.0	0.9	−0.2	1.0	1.7
−8.3	−2.1	−6.8	−4.3	2.2	−7.8	2.9	−0.6	1.1	−0.1	−0.4	0.3
−3.1	−1.0	−6.5	−1.8	−5.2	−5.7	1.2	−1.6	−0.1	−1.5	0.7	−0.4
−7.0	−2.2	−6.5	−1.0	−3.0	−3.6	2.2	−0.4	−2.2	−0.1		
−5.2	−2.0	−2.1	−5.6	−4.4	−3.3						
−4.0	−4.9	−4.7	−3.8	−5.9	−2.5						
−0.3	−6.2	−6.8	1.7	0.3	−2.3						
0.4	−5.3	0.2	−2.2	−5.1							

Select **Graph ➤ Stem-and-Leaf** from the menu and use the change in mineral content for the Graph variable. Enter the column containing the group for the By variable. Minitab will produce a separate plot for each group. Also, select **Stat ➤ Basic Statistics ➤ Display Descriptive Statistics** to compare the two distributions. Do the data show distinctly greater bone mineral loss among the breast-feeding women?

2.20 In 1798 the English scientist Henry Cavendish measured the density of the earth with great care. It is common practice to repeat careful measurements several times and use the mean as the final result. Cavendish repeated his work 29 times. Here and in EX02-20.MTW are his results (the data give the density of the earth as a multiple of the density of water):

5.50	5.61	4.88	5.07	5.26	5.55	5.36	5.29	5.58	5.65
5.57	5.53	5.62	5.29	5.44	5.34	5.79	5.10	5.27	5.39
5.42	5.47	5.63	5.34	5.46	5.30	5.75	5.68	5.85	

Present these measurements with a graph of your choice. Scientists usually give the mean and standard deviation to summarize a set of measurements. Does the shape of this distribution suggest that $\bar{x}$ and s are adequate summaries? Calculate $\bar{x}$ and s.

2.21 You are planning a sample survey of households in California. You decide to select households separately within each county and to choose more households from the more populous counties. To aid in the planning, Table 2.3 in BPS and TA02-03.MTW give the populations of California counties from the 2000 census. Examine the distribution of county populations both graphically and numerically, using whatever tools are most suitable. Write a brief description of the main features of this distribution. Sample surveys often select households from all of the most populous counties but from only some of the less populous. How would you divide California counties into three groups according to population, with the intent of including all of the first group, half of the second, and a smaller fraction of the third in your survey?

2.22 The University of Miami Hurricanes have been among the more successful teams in college football. Table 2.4 gives the weights in pounds and positions of the players on the 2002 team. The positions are quarterback

(QB), running back (RB), offensive line (OL), wide receiver (WR), tight end (TE), kicker/punter (KP), defensive back (DB), linebacker (LB), and defensive line (DL).

(a) Select **Graph ➤ Boxplots** to make side-by-side boxplots of the weights for running backs, wide receivers, offensive linemen, defensive linemen, linebackers, and defensive backs.

(b) Briefly compare the weight distributions. Which position has the heaviest players overall? Which has the lightest?

(c) Are any individual players outliers within their position?

2.23 Guinea pig survival times. Here are the survival times in days of 72 guinea pigs after they were injected with infectious bacteria in a medical experiment. Survival times, whether of machines under stress or cancer patients after treatment, usually have distributions that are skewed to the right.

43	45	53	56	56	57	58	66	67	73	74	79
80	80	81	81	81	82	83	83	84	88	89	91
91	92	92	97	99	99	100	100	101	102	102	102
103	104	107	108	109	113	114	118	121	123	126	128
137	138	139	144	145	147	156	162	174	178	179	184
191	198	211	214	243	249	329	380	403	511	522	598

(a) Graph the distribution and describe its main features. Does it show the expected right skew?

(b) Which numerical summary would you choose for these data? Select **Stat ➤ Basic Statistics ➤ Display Descriptive Statistics** to calculate your chosen summary. How does it reflect the skewness of the distribution?

2.25 In 2002, the Chicago Cubs failed once again to reach the National League playoffs. Table 2.5 in BPS and TA02-05.MTW give the salaries of the Cubs' players as of opening day of the season. Describe the distribution of salaries both with a graph and with a numerical summary. Then write a brief description of the important features of the distribution.

Chapter 3
Normal Distributions

Topic to be covered in this chapter:

Normal Calculations

Normal Calculations

Sometimes the Normal density can describe the overall pattern of a distribution. Figure 3.1 of BPS is a histogram of the scores of all 947 seventh-grade students in Gary, Indiana, on the vocabulary part of the Iowa Test of Basic Skills. To see how well the Normal density describes these data, we can select **Graph ➤ Histogram** and select the With Fit option to obtain a histogram with a Normal curve fitted to the data.

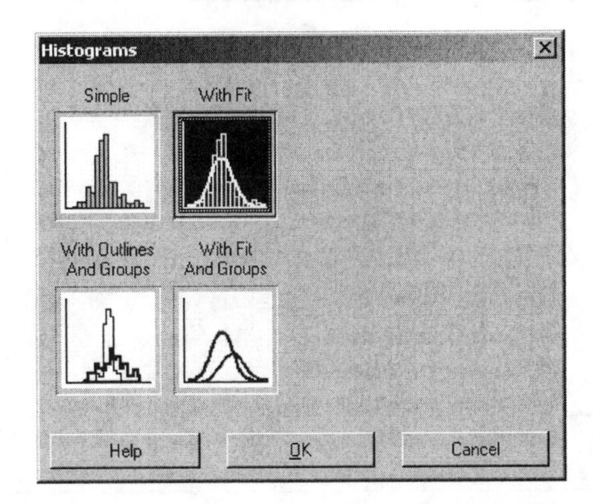

In the following dialog box, select the variable to be graphed and click OK. Minitab will produce a histogram with a Normal density curve for the variable. The Normal density curve selected will be based on the data's mean and standard deviation.

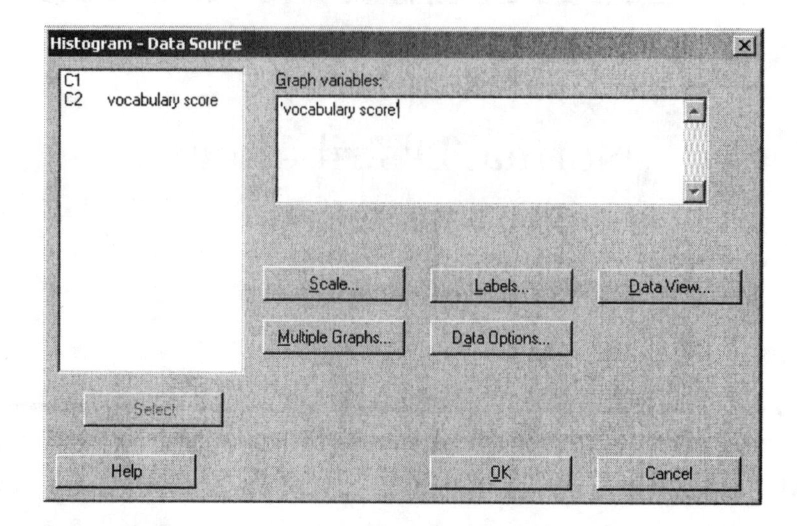

The smooth curve drawn through the histogram is a good description of the overall pattern of the data. It will be easier to use the density curve instead of the histogram for some calculations.

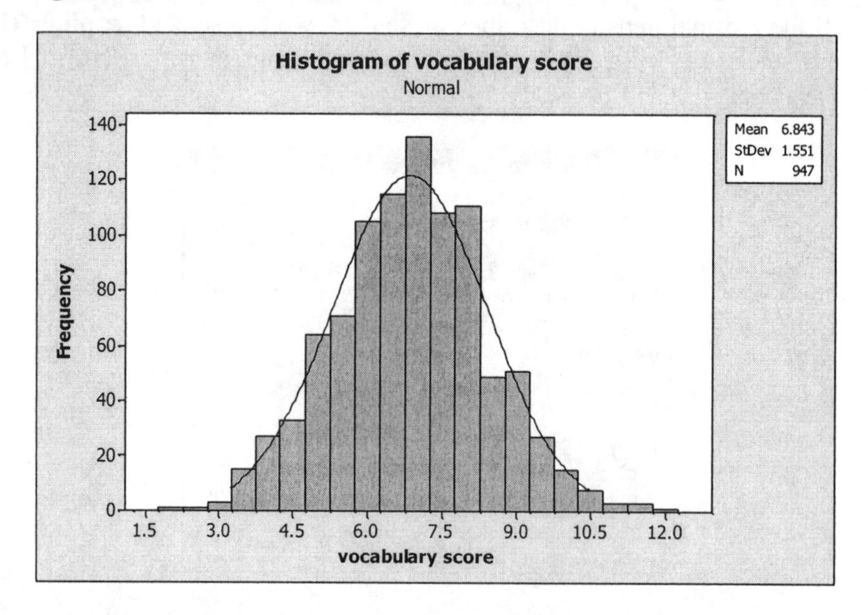

Minitab can be used to perform Normal distribution calculations. If data in a column are Normally distributed, then the data can be standardized to obtain data with a standard Normal distribution, that is, those with mean equal to zero and standard deviation equal to one. Select

Calc ➤ Standardize

from the menu to standardize. By selecting "Subtract mean and divide by std. dev." the command calculated the standardized values, $z = (x - \overline{x})/s$. The results can be stored in C2.

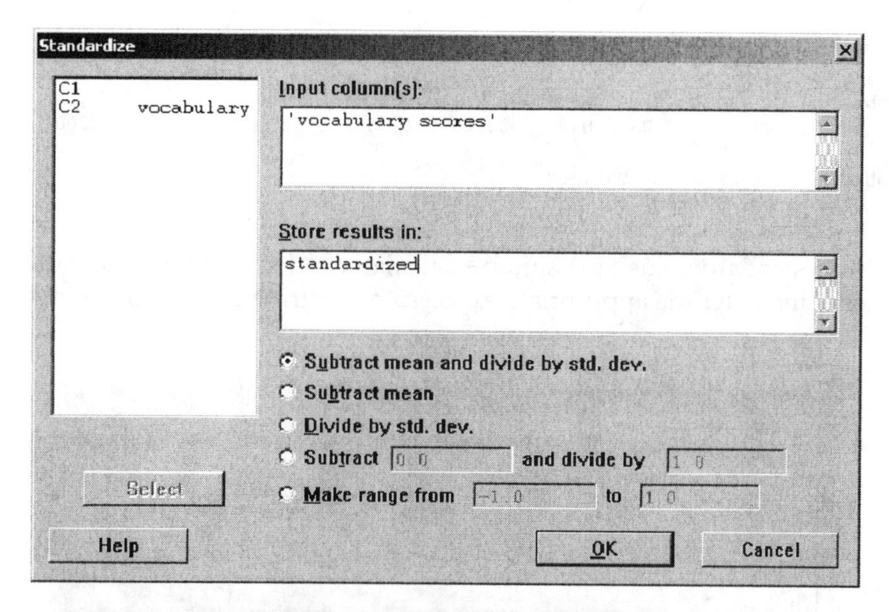

Worksheet 1 ***

↓	C1	C2	C3	C4	C5	C6	C7
		vocabulary scores	standardized				
1	5.4	5.8	-0.67213				
2	6.8	6.3	-0.34981				
3	2.0	7.8	0.61713				
4	7.6	7.3	0.29481				
5	6.6	9.7	1.84192				
6	7.8	7.1	0.16589				
7	8.1	8.1	0.81051				
8	5.6	5.8	-0.67213				
9	6.0	8.4	1.00390				
10	7.9	8.4	1.00390				
11	5.1	3.6	-2.09031				

The standardized vocabulary scores (or z-score) will tell how far above or below the mean a particular car falls. The measure is in units of standard deviations. The first student has a score of 5.8, a value that is below the mean ($z = -0.67213$). The second score, 6.3, is also below the mean, but the third, 7.8, is above the mean ($z = 0.61713$).

We could examine the standardized values to see how well they obey the 69-95-99.7 rule. Approximately 68% of the standardized values should have values between −1 and +1, 95% should have values between −2 and +2, and 99.7% should have values between −3 and +3.

Another way to obtain standardized values (z-scores) is to use Minitab's calculator. From the following descriptive information, we can standardize the vocabulary scores by subtracting the mean (6.8427) and dividing by the standard deviation (1.5513).

Descriptive Statistics: vocabulary scores

```
Variable            N  N*    Mean  SE Mean   StDev  Minimum       Q1  Median
vocabulary score  947   0  6.8427   0.0504  1.5513   2.1000   5.8000  6.8000

Variable               Q3  Maximum
vocabulary score   7.9000  12.1000
```

To standardize using Minitab's calculator, select **Calc ➤ Calculator** from the menu and enter the appropriate expression illustrated in the following dialog box.

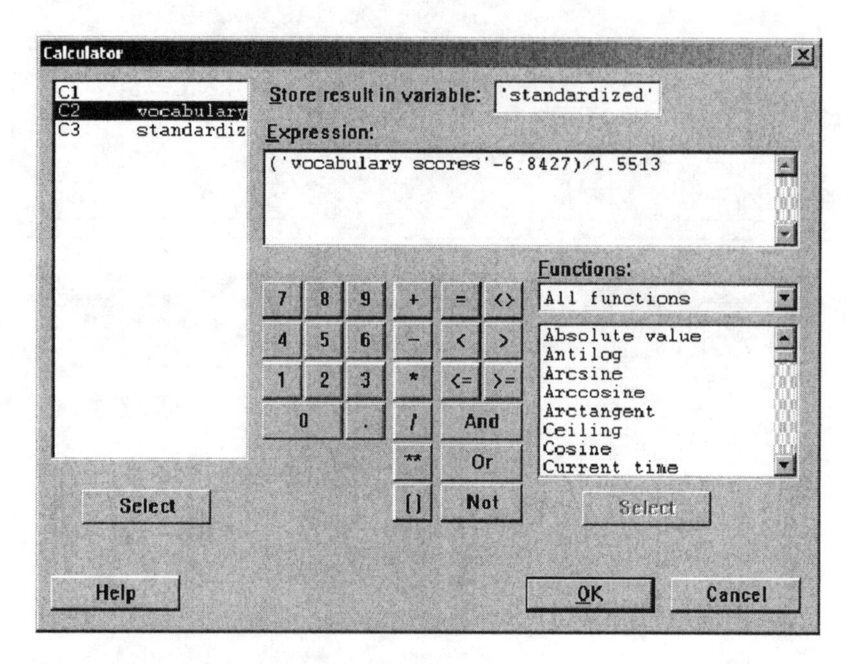

You can use Minitab to do probability calculations for the Normal distribution by selecting

Calc ➤ Probability Distributions ➤ Normal

from the menu. Both forward and backward probabilities can be calculated.

Example 3.4 in BPS concerns the heights of young women. The heights are approximately Normal with a mean of about 64 inches and a standard deviation of 2.7 inches. To find the proportion of women who are less than 70 inches tall, we select **Calc ➤ Probability Distributions ➤ Normal** from the menu. In the dialog box, we select Cumulative probability, fill in the Mean, Standard deviation, and Input constant boxes as follows.

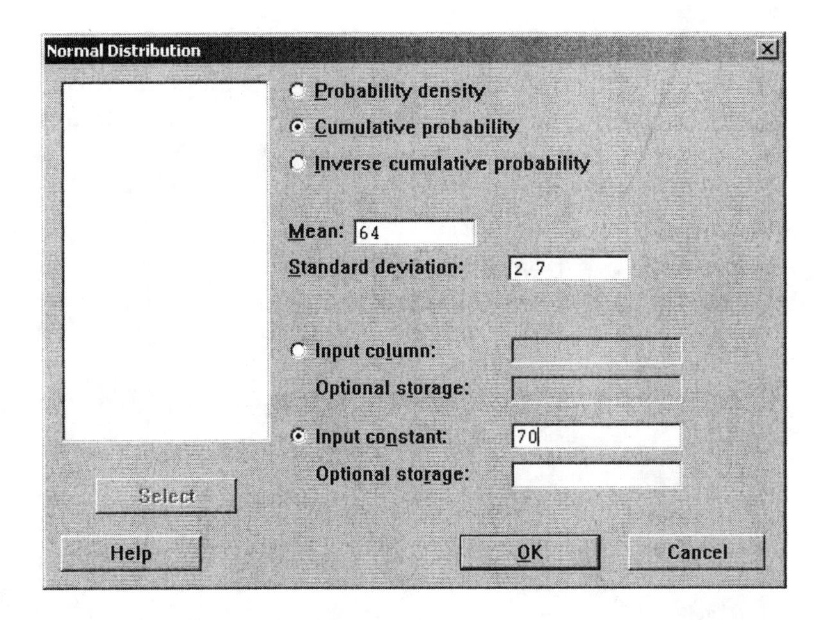

After clicking on OK, we obtain the following information in Minitab's Session window. The result says that the proportion of women who are less than 70 inches tall is .986866. This is slightly different from the result that would be obtained using Table A since it is not required to round the standardized value.

Cumulative Distribution Function

```
Normal with mean = 64 and standard deviation = 2.7

 x  P( X <= x )
70      0.986866
```

We can also use Minitab to do backward calculations. Example 3.8 of BPS shows scores on the SAT verbal test in 2002. The scores follow approximately the N(504, 110) distribution and we'd like to find how high a student must score to place in the top 10% of all students taking the SAT. Again, we select **Calc ➤ Probability Distributions ➤ Normal** from the menu. Since we are doing a backward calculation, we check Inverse cumulative probability. We fill in the Mean, Standard deviation, and Input constant boxes as follows. Notice that since we want the value for the top 10%, the input constant is 0.9 corresponding to 90% below the calculated value.

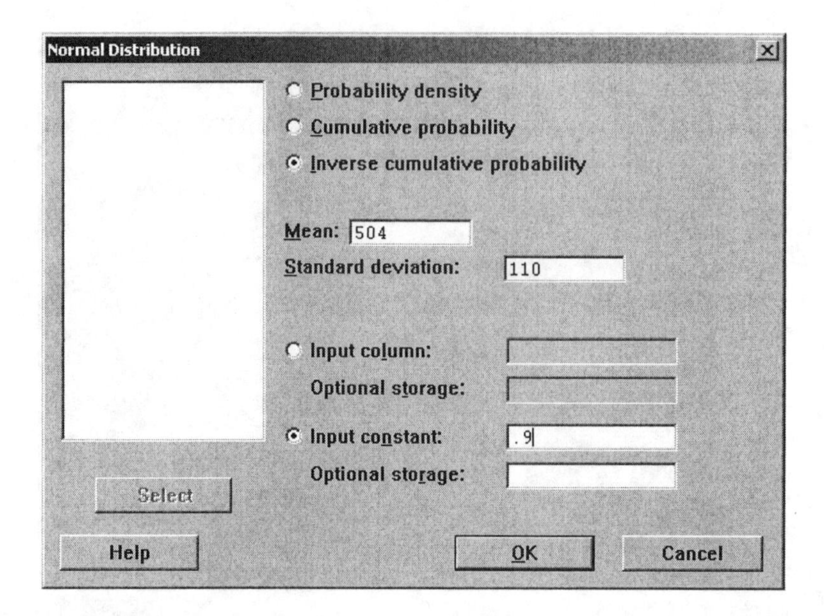

After clicking on OK in the dialog box, we obtain the following results in the Session window. We see that a student must score above 646.252 (i.e., 647 or above) to be in the top 10%.

Inverse Cumulative Distribution Function

```
Normal with mean = 504 and standard deviation = 111

P( X <= x )        x
        0.9  646.252
```

EXERCISES

3.10 Select **Calc ➤ Probability Distributions ➤ Normal** from the menu to find the proportion of observations from a standard Normal distribution that satisfies each of the following statements. In each case, sketch a standard Normal curve and shade the area under the curve that is the answer to the question.

(a) $z < 2.85$

(b) $z > 2.85$

(c) z ≥ ◎1.66

(d) ◎1.66 < z < 2.85

3.11 How hard do locomotives pull? An important measure of the performance of a locomotive is its "adhesion," which is the locomotive's pulling force as a multiple of its weight. The adhesion of one 4400-horsepower diesel locomotive model varies in actual use according to a Normal distribution with mean $\mu = 0.37$ and standard deviation $\sigma = 0.04$. Select **Calc ➤ Probability Distributions ➤ Normal** from the menu to answer the following questions.

(a) What proportion of adhesions measured in use are higher than 0.40?

(b) What proportion of adhesions are between 0.40 and 0.50?

(c) Improvements in the locomotive's computer controls change the distribution of adhesion to a Normal distribution with mean μ = 0.41 and standard deviation σ = 0.02. Find the proportions in (a) and (b) after this improvement.

3.13 Scores on the Wechsler Adult Intelligence Scale (WAIS) are approximately Normally distributed with μ = 100 and σ = 15. Select **Calc ➤ Probability Distributions ➤ Normal** from the menu to answer the following questions.

(a) What IQ scores fall in the lowest 25% of the distribution?

(b) How high an IQ score is needed to be in the highest 5%?

3.15 Scores on the WAIS are approximately Normally distributed with μ = 100 and σ = 15. Select **Calc ➤ Probability Distributions ➤ Normal** from the menu to find about what percent of people have WAIS scores

(a) above 100?

(b) above 145?

(c) below 85?

3.16 People with WAIS scores below 70 are considered mentally retarded when, for example, applying for Social Security disability benefits. Select **Calc ➤ Probability Distributions ➤ Normal** from the menu to find the percent of adults who are retarded by this criterion?

3.17 The organization MENSA, which calls itself "the high IQ society," requires a WAIS score of 130 or higher for membership. (Similar scores on other tests are also accepted.) Select **Calc ➤ Probability Distributions ➤ Normal** from the menu to find the percent of adults would qualify for membership?

3.18 Select **Calc ➤ Probability Distributions ➤ Normal** from the menu to find the proportion of observations from a standard Normal distribution that falls in each of the following regions. In each case, sketch a standard Normal curve and shade the area representing the region.

(a) $z \leq -2.25$

(b) $z \geq -2.25$

(c) $z > 1.77$

(d) $-2.25 < z < 1.77$

3.19 Select **Calc ➤ Probability Distributions ➤ Normal** from the menu to find the value z of a standard Normal variable that satisfies each of the following conditions. In each case, sketch a standard Normal curve with your value of z marked on the axis.

(a) The point z with the proportion of observations that are less than z in a standard Normal distribution is 0.8.

(b) The point z such that 35% of all observations from a standard Normal distribution are greater than z.

3.20 The National Collegiate Athletic Association (NCAA) requires Division I athletes to score at least 820 on the combined mathematics and verbal parts of the SAT exam to compete in their first college year. (Higher scores are required for students with poor high school grades.) In 2002, the scores of the 1.3 million students taking the SATs were approximately Normal with mean 1020 and standard deviation 207. Select **Calc ➤ Probability Distributions ➤ Normal** from the menu to find the percent of all students who score less than 820.

3.21 The NCAA considers a student a "partial qualifier" eligible to practice and receive an athletic scholarship, but not to compete, if the combined SAT score is at least 720. Use the information in the previous exercise and select **Calc ➤ Probability Distributions ➤ Normal** from the menu to find the percent of all SAT scores that are less than 720.

3.22 The heights of women aged 20 to 29 follow approximately the $N(64, 2.7)$ distribution. Men the same age have heights distributed as N(69.3, 2.8). Select **Calc ➤ Probability Distributions ➤ Normal** from the menu to find the percent of young women who are taller than the mean height of young men.

3.23 The heights of women aged 20 to 29 follow approximately the $N(64, 2.7)$ distribution. Men the same age have heights distributed as N(69.3, 2.8). Select **Calc ➤ Probability Distributions ➤ Normal** from the menu to find the percent of young men who are shorter than the mean height of young women.

3.24 The length of human pregnancies from conception to birth varies according to a distribution that is approximately Normal with mean 266 days and standard deviation of 16 days. Select **Calc ➤ Probability Distributions ➤ Normal** from the menu to answer the following questions.

(a) What percent of pregnancies last less than 240 days (that's about 8 months)?

(b) What percent of pregnancies last between 240 and 270 days (roughly between 8 months and 9 months)?

(c) How long do the longest 20% of pregnancies last?

3.25 Changing the mean of a Normal distribution by a moderate amount can greatly change the percent of observations in the tails. Suppose that a college is looking for applicants with SAT math scores 750 and above. Select **Calc ➤ Probability Distributions ➤ Normal** from the menu to answer the following questions.

(a) In 2002, the scores of men on the math SAT followed the $N(534, 116)$ distribution. What percent of men scored 750 or better?

(b) Women's SAT math scores that year had the $N(500, 110)$ distribution. What percent of women scored 750 or better? You see that the percent of men above 750 is almost three times the percent of women with such high scores.

3.28 The median of any Normal distribution is the same as its mean. We can use Normal calculations to find the quartiles for Normal distributions. Select **Calc ➤ Probability Distributions ➤ Normal** from the menu to answer the following questions.

(a) What is the area under the standard Normal curve to the left of the first quartile? Use this to find the value of the first quartile for a standard Normal distribution. Find the third quartile similarly.

(b) Your work in (a) gives the z-scores for the quartiles of any Normal distribution. What are the quartiles for the lengths of human pregnancies? (Use the distribution in Exercise 3.24.)

3.29 The *deciles* of any distribution are the points that mark off the lowest 10% and the highest 10%. On a density curve, these are the points with areas 0.1 and 0.9 to their left under the curve. Select **Calc ➤ Probability Distributions ➤ Normal** from the menu to answer the following questions.

(a) What are the deciles of the standard Normal distribution?

(b) The heights of young women are approximately Normal with mean 64 inches and standard deviation 2.7 inches. What are the deciles of this distribution?

Chapter 4
Scatterplots and Correlation

Topics to be covered in this chapter:

Scatterplots
Correlation

Scatterplots

Often we are interested in illustrating the relationships between two variables, such as the relationship between height and weight, between smoking and lung cancer, or between advertising expenditures and sales. If both variables are quantitative, the most useful display of their relationship is the scatterplot. Scatterplots can be produced by selecting

Graph ➤ Scatterplot

from the menu. Choose Simple and click OK in the following dialog box.

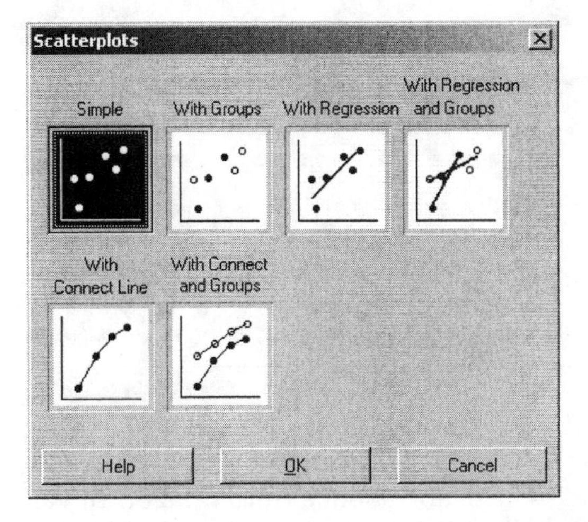

For illustration, we will consider the relationship between the percent of a state's high school graduates who took the SAT in 2002 and the state average SAT verbal score that year. The data are stored in EG04-03.MTW. We think that "percent taking" will help explain "average score." Therefore, "percent taking" is the explanatory variable and "average score" is the response variable. We want to see how average score changes when percent taking changes, so we put percent taking (the explanatory variable) on the horizontal axis. In the dialog box,

"SATV" was selected as the response (Y) variable and "PctS" as the explanatory (X) variable. Clicking on OK will produce a scatterplot.

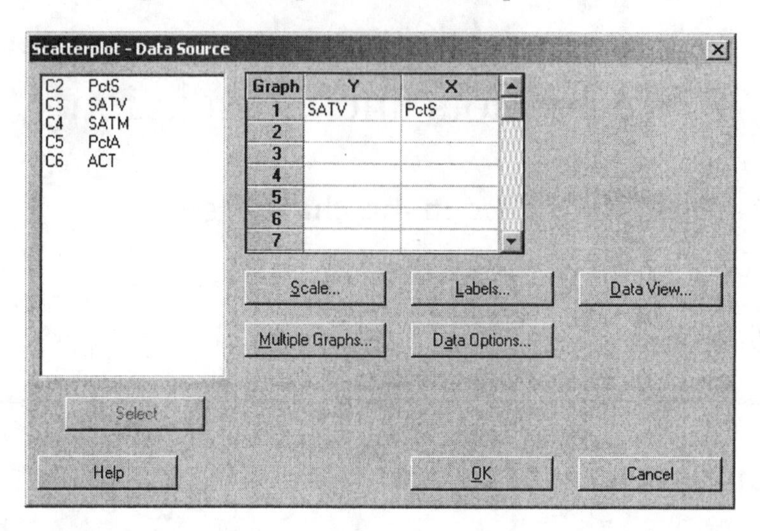

The scatterplot will appear as follows. There are two distinct clusters of states with a gap between them. There are no clear outliers. That is, no points fall clearly outside the clusters.

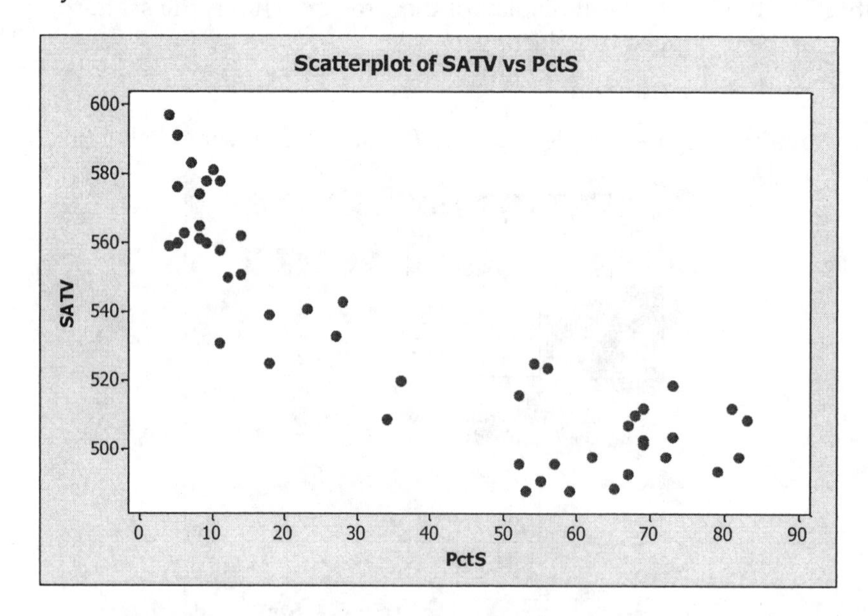

We can add information about a third categorical variable to a scatterplot by using different symbols for different points. We added a variable "Southern" identifying the 13 southern states. Southern states are marked with a "1" and other states are marked with a "0" in this new column.

A labeled scatterplot can then be obtained by selecting **Graph ➤ Scatterplot** from the menu and choosing With Groups in the dialog box shown on page 41. We now enter the Y variable, the X variable, and a categorical variable as follows.

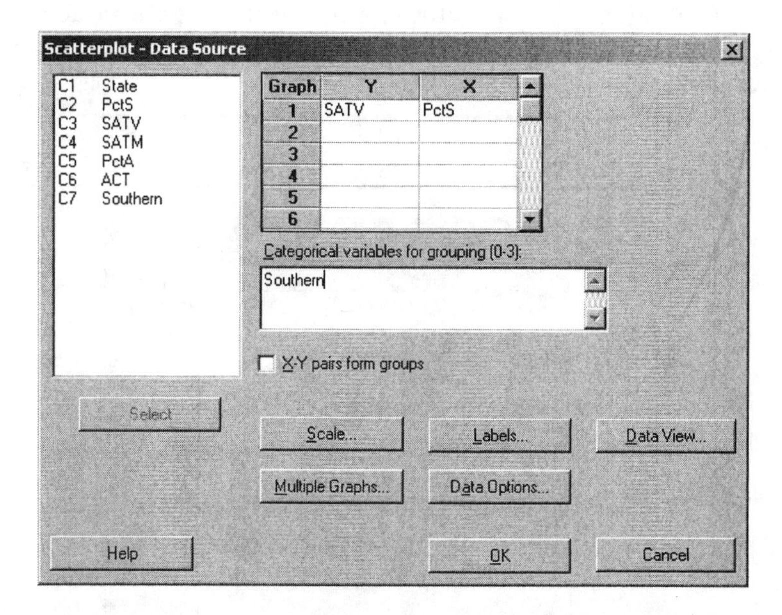

After clicking OK, we obtain a scatterplot with the Southern states plotted with a different plot symbol. We see from the scatterplot on the following page that the southern states blend in with the rest of the country.

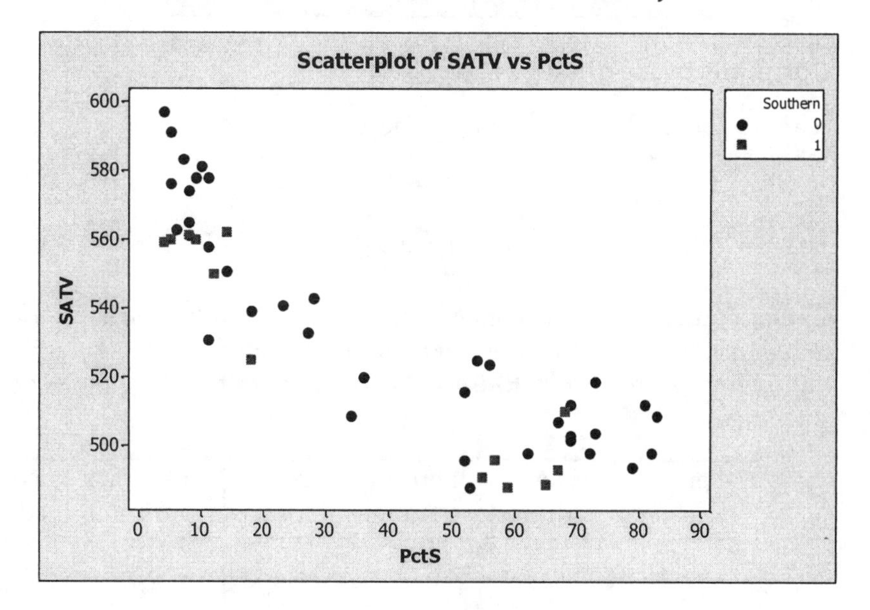

Correlation

We can compute the correlation coefficient between two quantitative variables using Minitab. The correlation coefficient can be calculated by selecting

Stat ➤ Basic Statistics ➤ Correlation

and selecting the desired variables. If more than two variables are specified, Minitab will print a table giving the correlation coefficients between all pairs of columns.

A correlation calculation with data for all three 2001-model four-wheel-drive minivans is illustrated below. The data represent the city miles per gallon and highway miles per gallon for all three minivans.

City MPG	18.6	19.2	18.8
Highway MPG	28.7	29.3	29.2

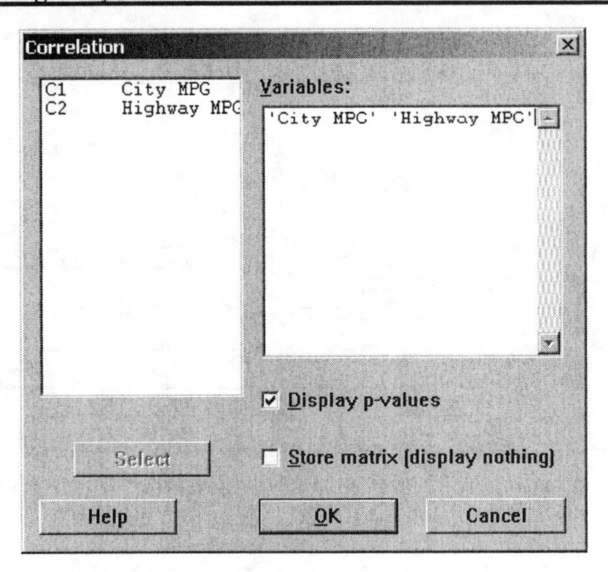

Correlations: Highway, City

```
Pearson correlation of Highway and City = 0.849
P-Value = 0.355
```

EXERCISES

4.4 One of nature's patterns connects the percent of adult birds in a colony that return from the previous year and the number of new adults that join the colony. Here and in EX04-04.MTW are data for 13 colonies of sparrowhawks.

Percent returning	New adults	Percent returning	New adults	Percent returning	New adults
74	5	62	15	46	18
66	6	52	16	60	19
81	8	45	17	46	20
52	11	62	18	38	20
73	12				

(a) Select **Graph ➤ Scatterplot** from the menu to plot the count of new birds (response) against the percent of returning birds (explanatory).

(b) Describe the form, direction, and strength of the relationship between number of new sparrowhawks in a colony and percent of returning adults, as displayed in your plot.

For short-lived birds, the association between these variables is positive: changes in weather and food supply drive the populations of new and returning birds up or down together. For long-lived territorial birds, on the other hand, the association is negative because returning birds claim their territories in the colony and don't leave room for new recruits. Which type of species is the sparrowhawk?

4.6 How does the fuel consumption of a car change as its speed increases? Here are data for a British Ford Escort. Speed is measured in kilometers per hour, and fuel consumption is measured in liters of gasoline used per 100 kilometers traveled.

Speed (km/h)	Fuel used (liters/100 km)	Speed (km/h)	Fuel used (liters/100 km)
10	21.00	90	7.57
20	13.00	100	8.27
30	10.00	110	9.03
40	8.00	120	9.87
50	7.00	130	10.79
60	5.90	140	11.77
70	6.30	150	12.83
80	6.95		

(a) Select **Graph ➤ Scatterplot** from the menu to make a scatterplot. (Which is the explanatory variable?)

(b) Describe the form of the relationship. It is not linear. Explain why the form of the relationship makes sense.

(c) It does not make sense to describe the variables as either positively associated or negatively associated. Why?

(d) Is the relationship reasonably strong or quite weak? Explain your answer.

4.7 How fast do icicles grow? Japanese researchers measured the growth of icicles in a cold chamber under various conditions of temperature, wind, and water flow. Table 4.2 and TA04-02.MTW contain data produced under two sets of conditions. In both cases, there was no wind and the temperature was set at −11°C. Water flowed over the icicle at a higher rate (29.6 milligrams per second) in run 8905 and at a slower rate (11.9 mg/s) in run 8903.

(a) Select **Graph ➤ Scatterplot** from the menu to make a scatterplot of the length of the icicle in centimeters versus time in minutes, using separate symbols for the two runs.

(b) What do your plots show about the pattern of growth of icicles? What do they show about the effect of changing the rate of water flow on icicle growth?

4.8 Archaeopteryx is an extinct beast having feathers like a bird but teeth and a long bony tail like a reptile. Only six fossil specimens are known. Because these specimens differ greatly in size, some scientists think they belong to different species. We will examine some data. If the specimens belong to the same species and differ in size because some are younger than others, there should be a positive linear relationship between the lengths of a pair of bones from all individuals. An outlier from this relationship would suggest a different species. Here are data on the lengths in centimeters of the femur (a thigh bone) and the humerus (a bone in the upper arm) for the five specimens of which both bones are preserved:

Femur	38	56	59	64	74
Humerus	41	63	70	72	84

(a) Select **Graph ➤ Scatterplot** from the menu to make a scatterplot. Do you think that all five specimens come from the same species?

(b) Select **Stat ➤ Basic Statistics ➤ Correlation** to find the correlation.

4.9 Changing the correlation.

(a) Select **Stat ➤ Basic Statistics ➤ Correlation** to find the correlation between the percent of returning birds and the number of new birds from the data in Exercise 4.4 and EX04-04.

(b) Select **Graph ➤ Scatterplot** from the menu to make a scatterplot of the data with two new points added: point A, 10% return, 25 new birds, point B, 40% return, 5 new birds. Find two new correlations, for the original data plus point A, and for the original data plus point B.

(c) Explain in terms of what correlation measures why adding Point A makes the correlation stronger (closer to –1) and adding Point B makes the correlation weaker (closer to 0).

4.11 The gas mileage of an automobile first increases and then decreases as the speed increases. Suppose that this relationship is very regular, as shown by the following data on speed (miles per hour) and mileage (miles per gallon):

Speed	20	30	40	50	60
MPG	24	28	30	28	24

Select **Graph ➤ Scatterplot** from the menu to make a scatterplot of mileage versus speed. Select **Stat ➤ Basic Statistics ➤ Correlation** to find the correlation between speed and mileage. Explain why the correlation is 0 even though there is a strong relationship between speed and mileage.

4.12 Table 1.2 in BPS and TA01-02.MTW give the city and highway gas mileages for two-seater cars. We expect a positive association between the city and highway mileages of a group of vehicles. Select **Graph ➤ Scatterplot** from the menu to make a scatterplot that shows the relationship between city and highway mileage, using city mileage as the explanatory variable. Describe the overall pattern. Does the outlier (the Honda Insight) extend the pattern of the other cars or is it far from the line they form?

4.14 Select **Stat ➤ Basic Statistics ➤ Correlation** to find the correlation between city and highway mileage in Table 1.2 and TA01-02.MTW, leaving out the Insight. Explain how the correlation matches the pattern of the plot. Based on your plot in Exercise 4.12, will adding the Insight make the correlation stronger (closer to 1) or weaker? Verify your guess by calculating the correlation for all 22 cars.

4.16 The Sanchez household is about to install solar panels to reduce the cost of heating their house. In order to know how much the solar panels help, they record their consumption of natural gas before the panels are installed. Gas consumption is higher in cold weather, so the relationship between outside temperature and gas consumption is important. Here and in EX04-16.MTW are data for 16 consecutive months.

Month	Nov.	Dec.	Jan.	Feb.	Mar.	Apr.	May	June
Degree-days	24	51	43	33	26	13	4	0
Gas used	6.3	10.9	8.9	7.5	5.3	4.0	1.7	1.2
Month	July	Aug.	Sept.	Oct.	Nov.	Dec.	Jan.	Feb.
Degree-days	0	1	6	12	30	32	52	30
Gas used	1.2	1.2	2.1	3.1	6.4	7.2	11.0	6.9

Outside temperature is recorded in degree-days, a common measure of demand for heating. A day's degree-days is the number of degrees its average temperature falls below 65°F. Gas used is recorded in hundreds of cubic feet. Select **Graph ➤ Scatterplot** to make a plot and describe the pattern. Is correlation a reasonable way to describe the pattern? Select **Stat ➤ Basic Statistics ➤ Correlation** to find the correlation.

4.18 After the Sanchez household gathered the information recorded in Exercise 4.16, they added solar panels to their house. They then measured their natural gas consumption for 23 more months. Here and in EX04-18.MTW are the data.

Degree-days	19	3	3	0	0	0	8	11	27	46	38	34
Gas used	3.2	2.0	1.6	1.0	0.7	0.7	1.6	3.1	5.1	7.7	7.0	6.1
Degree-days	16	9	2	1	0	2	3	18	32	34	40	
Gas used	3.0	2.1	1.3	1.0	1.0	1.0	1.2	3.4	6.1	6.5	7.5	

Add the new data to EX04-16.MTW along with the extra column "solar panels" to indicate whether or not the data were recorded after the solar panels were added to their house. Select **Graph ➤ Scatterplot ➤ With Groups** to make a scatterplot with a different color and symbol for the new data. What do the before-and-after data show about the effect of solar panels?

4.20 Often the percent of an animal species in the wild that survive to breed again is lower following a successful breeding season. This is part of nature's self-regulation, tending to keep population size stable. A study of merlins (small falcons) in northern Sweden observed the number of breeding pairs in an isolated area and the percent of males (banded for identification) who returned the next breeding season. Here are data for nine years:

Breeding pair	Percent of males returning
28	82
29	83, 70, 61
30	69
32	58
33	43
38	50, 47

(a) Why is the response variable the percent of males that return rather than the number of males that return?

(b) Select **Graph ➤ Scatterplot** from the menu to make a scatterplot. Describe the pattern. Do the data support the theory that a smaller percent of birds survive following a successful breeding season?

4.21 Many mutual funds compare their performance with that of a benchmark, an index of the returns on all securities of the kind the fund buys. The Vanguard International Growth Fund, for example, takes as its benchmark the Morgan Stanley EAFE (Europe, Australasia, Far East) index of overseas stock market performance. Here and in EX04-21.MTW are the percent returns for the fund and for the EAFE from 1982 (the first full year of the fund's existence) to 2001:

Year	Fund	EAFE	Year	Fund	EAFE
1982	5.27	−1.86	1992	−5.79	−12.17
1983	43.08	23.69	1993	44.74	32.56
1984	−1.02	7.38	1994	0.76	7.78
1985	56.94	56.16	1995	14.89	11.21
1986	56.71	69.44	1996	14.65	6.05
1987	12.48	24.63	1997	4.12	1.78
1988	11.61	28.27	1998	16.93	20.00
1989	24.76	10.54	1999	26.34	26.96
1990	−12.05	−23.45	2000	−8.60	−14.17
1991	4.74	12.13	2001	−18.92	−21.44

Select **Graph ➤ Scatterplot** to make a scatterplot suitable for predicting fund returns from EAFE returns. Is there a clear straight-line pattern? How strong is this pattern? Select **Stat ➤ Basic Statistics ➤ Correlation** to give a numerical measure. Are there any extreme outliers from the straight-line pattern?

4.24 To demonstrate the effect of nematodes (microscopic worms) on plant growth, a student introduces different numbers of nematodes into 16 planting pots. He then transplants a tomato seedling into each pot. Here and in EX04-24.MTW are data on the increase in height of the seedlings (in centimeters) 14 days after planting:

Nematodes	Seedling growth			
0	10.8	9.1	13.5	9.2
1,000	11.1	11.1	8.2	11.3
5,000	5.4	4.6	7.4	5.0
10,000	5.8	5.3	3.2	7.5

(a) Select **Graph ➤ Scatterplot** to make a scatterplot of the response variable (growth) against the explanatory variable (nematode count). Then compute the mean growth for each group of seedlings, plot the means against the nematode counts, and connect these four points with line segments.

(b) Briefly describe the conclusions about the effects of nematodes on plant growth that these data suggest.

4.25 How many corn plants are too many? How much corn per acre should a farmer plant to obtain the highest yield? Too few plants will give a low yield. On the other hand, if there are too many plants, they will compete with each other for moisture and nutrients, and yields will fall. To find the best planting rate, plant at different rates on several plots of ground and measure the harvest. (Be sure to treat all the plots the same except for the planting rate.) Here and in EX04-25.MTW are data from such an experiment:

Plants per acre	Yield (bushels per acre)			
12,000	150.1	113.0	118.4	142.6
16,000	166.9	120.7	135.2	149.8
20,000	165.3	130.1	139.6	149.9
24,000	134.7	138.4	156.1	
28,000	119.0	150.5		

(a) Is yield or planting rate the explanatory variable?

(b) Select **Graph ➤ Scatterplot** from the menu to make a scatterplot of yield and planting rate.

(c) Describe the overall pattern of the relationship. Is it linear? Is there a positive or negative association or neither?

(d) Find the mean yield for each of the five planting rates. Plot each mean yield against its planting rate on your scatterplot and connect these five points with lines. This combination of numerical description and graphing makes the relationship clearer. What planting rate would you recommend to a farmer whose conditions were similar to those in the experiment?

4.27 Is wine good for your heart? There is some evidence that drinking moderate amounts of wine helps prevent heart attacks. Table 4.3 and TA04-03.MTW give data on yearly wine consumption (liters of alcohol from drinking wine, per person) and yearly deaths from heart disease (deaths per 100,000 people) in 19 developed nations.

(a) Select **Graph ➤ Scatterplot** from the menu to make a scatterplot that shows how national wine consumption helps explain heart disease death rates.

(b) Describe the form of the relationship. Is there a linear pattern? How strong is the relationship?

(c) Is the direction of the association positive or negative? Explain in simple language what this says about wine and heart disease. Do you think these data give good evidence that drinking wine causes a reduction in heart disease deaths? Why?

Chapter 5
Regression

Topics to be covered in this chapter:

Regression
Fitted Line Plots
Residual Plots

Regression

The scatterplot below shows that there is a linear relationship between the percent x of adult sparrowhawks that return to a colony from the previous year and the number y of new adult birds that join the colony. The scatterplot shows that the relationship is moderately strong. The data are given in Exercise 4.4 and EX04-04.MTW.

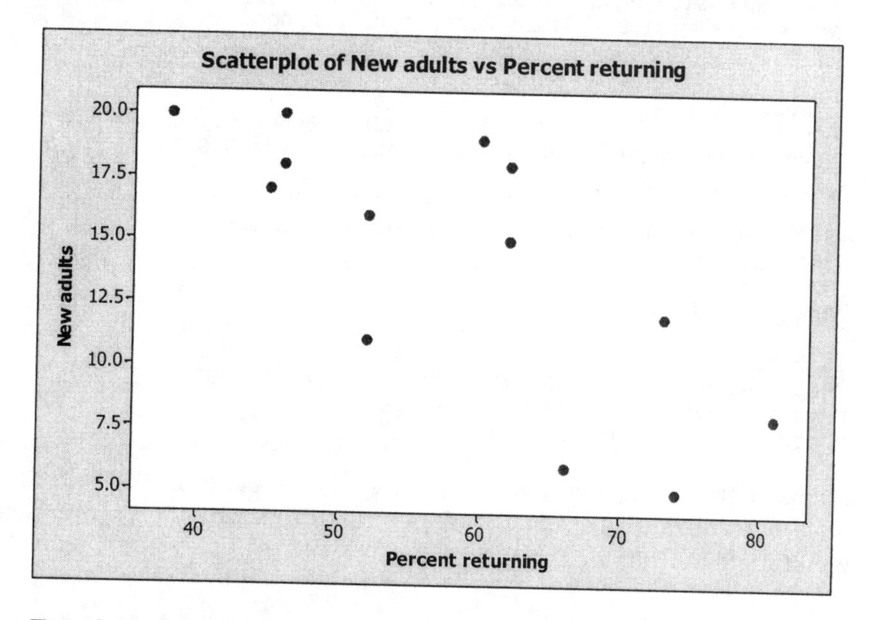

To calculate the least-squares line of the form $y = a + bx$ from data, select

Stat ➤ Regression ➤ Regression

from the menu. In the dialog box, enter "Percent returning" for the Predictor and "New adults" for the Response and click OK.

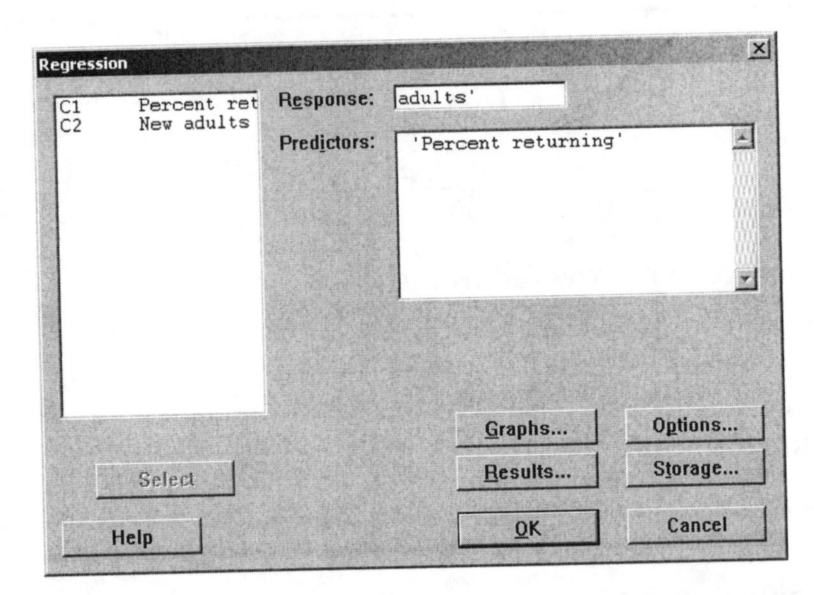

The following output gives the regression equation as $\hat{y} = 31.9 = 0.304x$. Under the Coef column, more accurate values for the intercept and slope are given as 31.934 and -0.30402, respectively.

Regression Analysis: New adults versus Percent returning

```
The regression equation is
New adults = 31.9 - 0.304 Percent returning


Predictor               Coef   SE Coef      T      P
Constant              31.934     4.838   6.60  0.000
Percent returning   -0.30402   0.08122  -3.74  0.003


S = 3.66689   R-Sq = 56.0%   R-Sq(adj) = 52.0%


Analysis of Variance

Source          DF      SS      MS      F      P
Regression       1  188.40  188.40  14.01  0.003
Residual Error  11  147.91   13.45
Total           12  336.31
```

The square of the correlation coefficient, r^2, also appears in the output. It is listed as a percentage (R-sq = 56%). Other useful information is provided and will be discussed in Chapter 21.

Fitted Line Plots

Fitted line plots can be obtained by selecting

Stat ➤ Regression ➤ Fitted line plot

from the menu and entering the appropriate predictor and response variable:

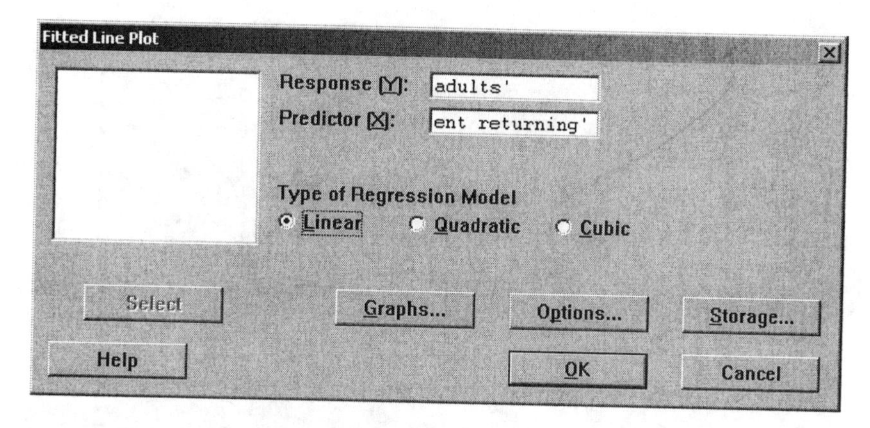

The fitted line plot shows the regression line plotted on the scatterplot. It also lists the equation for the regression line and the value for R-Sq:

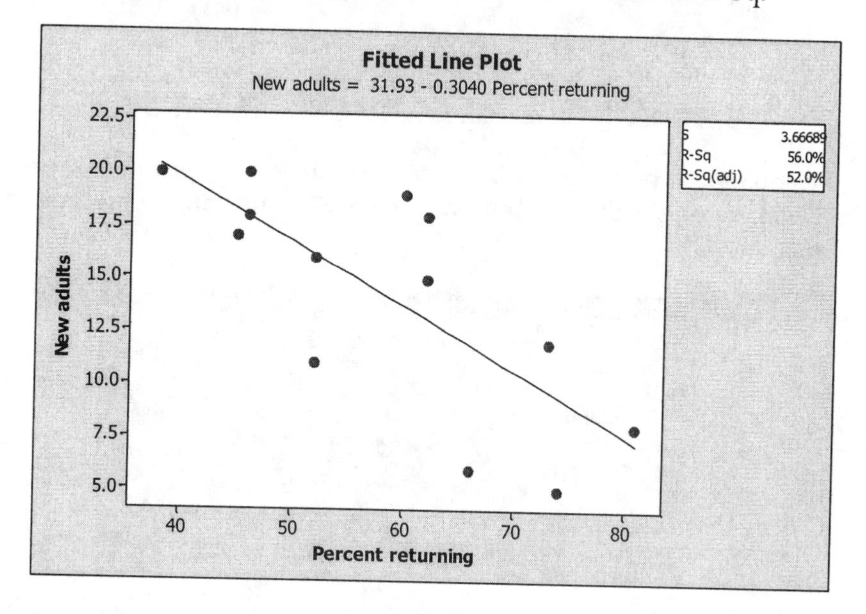

Sometimes we'd like to have a scatterplot with more than one regression line plotted. We can obtain one by selecting **Graph ➤ Scatterplot** from the menu and choosing With Regression and Groups in the dialog box. Another dialog box will appear in which you can select Y and X variables for graphing as well as a categorical variable for graphing.

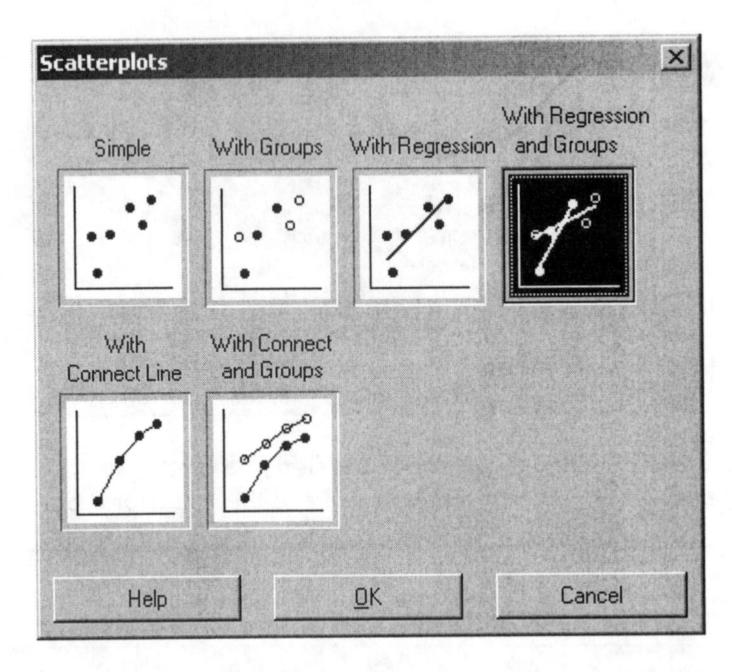

Residuals can be obtained by selecting **Stat ➤ Regression ➤ Regression** and then clicking on the Storage button. Check Residuals in the Storage subdialog box and click OK.

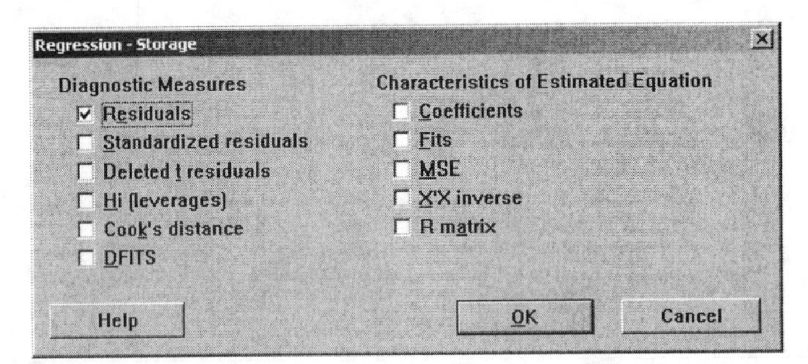

Residual Plots

To obtain a residual plot for regression, select **Stat ➤ Regression ➤ Regression** from the menu. Enter "Percent returning" for the Predictor and "New adults" for the Response, and click on the Graphs button. In the dialog box, enter "Percent returning" to obtain a plot of the residuals versus the explanatory variable and click OK.

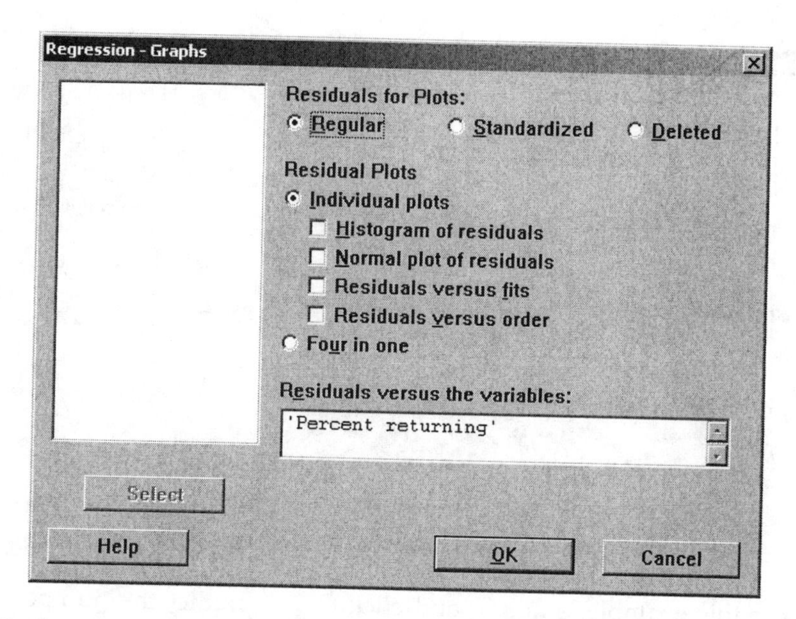

The residual plot that follows helps us assess the fit of a regression line. Since there is no pattern in the plot, the fit appears to be satisfactory.

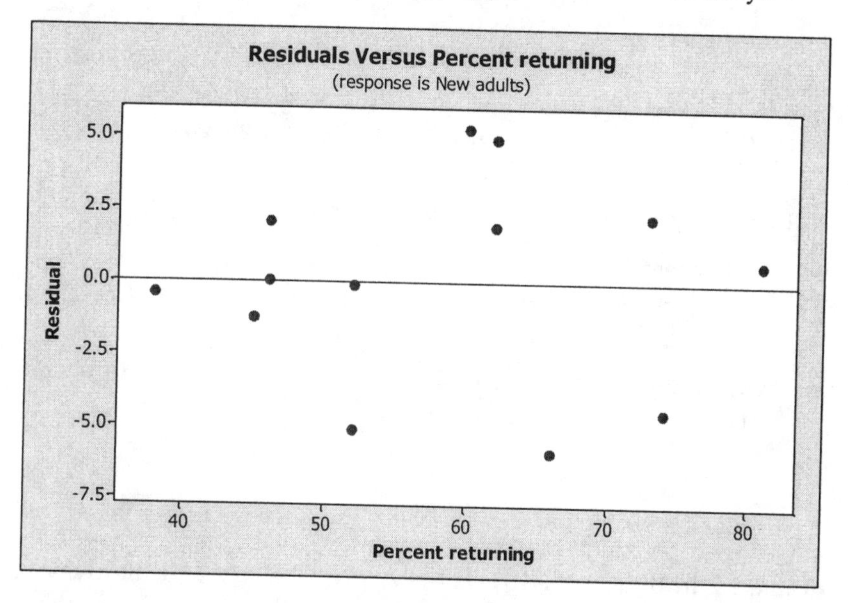

To use Minitab to make predictions based on a regression line, select **Stat ➤ Regression ➤ Regression** and click on the Options button. In the dialog box, enter the new value for which you would like to predict the response under Prediction intervals for new observations.

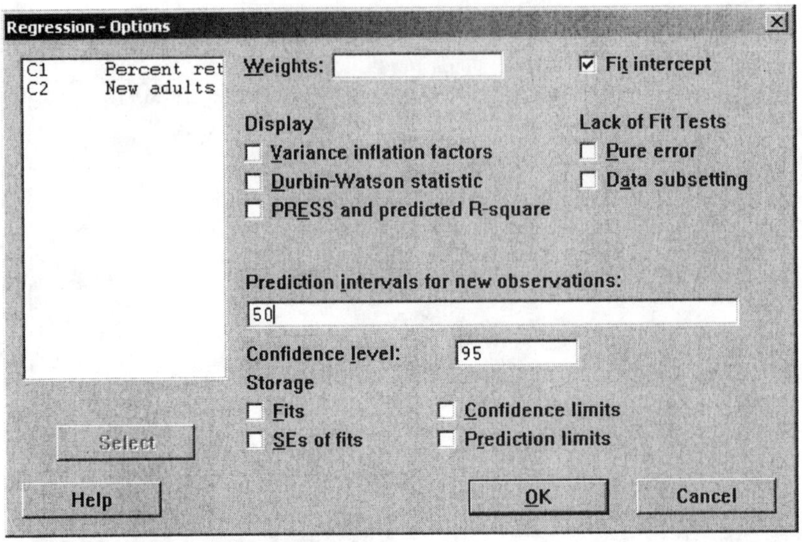

Regression - Options

C1 Percent ret
C2 New adults

Weights:

☑ Fit intercept

Display
☐ Variance inflation factors
☐ Durbin-Watson statistic
☐ PRESS and predicted R-square

Lack of Fit Tests
☐ Pure error
☐ Data subsetting

Prediction intervals for new observations:
50

Confidence level: 95

Storage
☐ Fits
☐ SEs of fits
☐ Confidence limits
☐ Prediction limits

Select

Help OK Cancel

For this example, enter 50 and click OK to predict the number of "New adults" if 50% of the adult sparrowhawks return to a colony. You will obtain the same regression output as shown previously, followed by the predicted values.

```
Predicted Values for New Observations

New
Obs    Fit  SE Fit        95% CI             95% PI
  1  16.73    1.22  (14.05, 19.41)  (8.23, 25.24)


Values of Predictors for New Observations

New    Percent
Obs  returning
  1       50.0
```

The value given in the column labeled Fit is the value of "New adults" given by the regression equation when "percent returning" is equal to 50.

EXERCISES

5.3 Table 1.2 in BPS and TA01-02.MTW give the city and highway gas mileages for two-seater cars. A scatterplot (Exercise 4.12) shows a strong positive linear relationship.

(a) Select **Stat ➤ Regression ➤ Regression** to find the least-squares regression line for predicting highway mileage from city mileage, using data from all 22 car models. Select **Stat ➤ Regression ➤ Fitted Line Plot** to make a scatterplot with the regression line.

(b) What is the slope of the regression line? Explain in words what the slope says about gas mileage for two-seater cars.

(c) Another two-seater is rated at 20 miles per gallon in the city. Predict its highway mileage.

5.4 Exercise 4.25 and EX04-25.MTW give data from an agricultural experiment. The purpose of the study was to see how the yield of corn changes with change in the planting rate (plants per acre).

(a) Select **Stat ➤ Regression ➤ Fitted Line Plot** to make a scatterplot of the data with the the least-squares regression line for predicting yield from planting rate added to the plot. Why should we not use regression for prediction in this setting?

(b) What is r^2? What does this value say about the success of the regression in predicting yield?

(c) Even regression lines that make no practical sense obey Facts 1 to 4 in BPS. Use the equation of the regression line you found in (a) to show when x is the mean planting rate, the predicted yield $\hat{y}$ is the mean of the observed yields.

5.5 In Exercise 5.3 you found the least-squares regression line for predicting highway mileage from city mileage for the 22 two-seater car models in Table 1.2 in BPS and TA01-02.MTW. Select **Calc ➤ Column Statistics** to find the mean city mileage and the mean highway mileage for these cars. Select **Stat ➤ Regression ➤ Regression** and click on the Options button to predict the highway mileage for a car with city mileage equal to the mean for the group. Explain why you knew the answer before doing the prediction.

5.7 Exercise 4.6 and EX04-06.MTW give data on the fuel consumption y of a car at various speeds x. Fuel consumption is measured in liters of gasoline per 100 kilometers driven and speed is measured in kilometers per hour.

(a) Select **Stat ➤ Regression ➤ Fitted Line Plot** to make a scatterplot of the observations with the regression line on plot.

(b) Would you use the regression line to predict y from x? Explain your answer.

(c) Select **Stat ➤ Regression ➤ Regression** and click on the Storage button to save the residuals. Select **Calc ➤ Column Statistics** to check that the residuals have sum zero (up to roundoff error).

(d) Select **Stat ➤ Regression ➤ Regression** and click on the Graphs button to make a plot of the residuals against the values of x. Notice that the residuals show the same pattern about this line as the data points show about the regression line in the scatterplot in (a).

5.8 We have seen that Child 18 in the Gesell data in Table 5.1 of BPS and TA05-01.MTW is an influential observation. Now we will examine the effect of Child 19, who is also an outlier in Figure 5.5.

(a) Select **Stat ➤ Regression ➤ Fitted Line Plot** to make a scatterplot with the regression line on it for the data. Remove Child 19 and Select **Stat ➤ Regression ➤ Regression** to find the regression line for the data without Child 19. Find the least-squares regression line of Gesell score on age at first word, leaving out Child 19.

Draw the new regression line (by hand) on your fitted line plot. Would you call Child 19 very influential? Why?

(b) How did removing Child 19 change the r^2 for this regression? Explain why r^2 changes in this direction when you drop Child 19.

5.9 The data on gas mileage of two-seater cars Table 1.2 of BPS and TA01-02.MTW contain an outlier, the Honda Insight. When we predict highway mileage from city mileage, this point is an outlier in both the x and y directions. We wonder if it influences the least-squares line.

(a) Select **Stat ➤ Regression ➤ Fitted Line Plot** to make a scatterplot with the least-squares line from all 22 car models drawn on it.

(b) Select **Stat ➤ Regression ➤ Regression** to find the least-squares line when the Insight is left out of the calculation. Draw this line on your plot.

(c) Influence is a matter of degree, not a yes-or-no question. Use both regression lines to predict highway mileages for city mileages 10, 20, and 25 mpg. (These values span the range of car models other than the Insight.) Click on the Options button after you select **Stat ➤ Regression ➤ Regression** to make a prediction. Do you think the Insight changes the predictions enough to be important to a car buyer?

5.10 The number of people living on American farms has declined steadily during the 20th century. Here and in EX05-10.MTW are data on the farm population (millions of persons) from 1935 to 1980.

Year	1935	1940	1945	1950	1955	1960	1965	1970	1975	1980
Population	32.1	30.5	24.4	23.0	19.1	15.6	12.4	9.7	8.9	7.2

(a) Select **Graph ➤ Plot** from the menu and make a plot of these data. Select **Stat ➤ Regression ➤ Regression** from the menu and find the least-squares regression line of farm population on year.

(b) According to the regression line, how much did the farm population decline each year on the average during this period? What percent of the observed variation in farm population is accounted for by linear change over time?

(c) Select **Stat ➤ Regression ➤ Regression** from the menu and use the Options button to predict the number of people living on farms in 1990. Is this result reasonable? Why?

5.15 How strongly do physical characteristics of sisters and brothers correlate? Here and in EX05-15.MTW are data on the height (in inches) of 11 adult pairs.

Brother	71	68	66	67	70	71	70	73	72	65	66
Sister	69	64	65	63	65	62	65	64	66	59	62

(a) Select **Stat ➤ Regression ➤ Regression** from the menu to find the least-squares line for predicting sister's height. What is the correlation between sister's height and brother's height?

(b) Damien is 70 inches tall. Predict the height of his sister Tonya.

5.20 Exercise 4.16 and EX04-6.MTW give data on degree-days and natural gas consumed by the Sanchez home for 16 consecutive months. There is a very strong linear relationship. Mr. Sanchez asks, "If a month averages 20 degree-days per day (that's 45°F), how much gas will we use?" Select **Stat ➤ Regression ➤ Regression** and click on the Options button to find the least-squares regression line and answer his question.

5.22 Exercise 4.20 and EX05-22.MTW give data on the number of breeding pairs of merlins in an isolated area in each of nine years and the percent of males who returned the next year. The data show that the percent returning is lower after successful breeding seasons and that the relationship is roughly linear. Select **Stat ➤ Regression ➤ Regression** and click on the Options button to find the least-squares regression line and predict the percent of returning males after a season with 30 breeding pairs.

5.23 Keeping water supplies clean requires regular measurement of levels of pollutants. The measurements are indirect—a typical analysis involves forming a dye by a chemical reaction with the dissolved pollutant, then passing light through the solution and measuring its "absorbence." To calibrate such measurements, the laboratory measures known standard solutions and uses regression to relate absorbence to pollutant concentration. This is usually done every day. Here and in EX05-23.MTW is one series of data on the absorbence for different levels of nitrates. Nitrates are measured in milligrams per liter of water.

Nitrates	50	50	100	200	400	800	1200	1600	2000	2000
Absorbence	7.0	7.5	12.8	24.0	47.0	93.0	138.0	183.0	230.0	226.0

(a) Chemical theory says that these data should lie on a straight line. If the correlation is not at least 0.997, something went wrong and the calibration procedure is repeated. Select **Graph ➤ Scatterplot** from the menu to plot the data. Select **Stat ➤ Basic Statistics ➤ Correlation** to find the correlation. Must the calibration be done again?

(b) What is the equation of the least-squares line for predicting absorbence from concentration? If the lab analyzed a specimen with 500 milligrams of nitrates per liter, what do you expect the absorbence to be? Select **Stat ➤ Regression ➤ Regression** and click on the Options button to find the prediction. Based on your plot and the correlation, do you expect your predicted absorbence to be very accurate?

5.25 Table 4.3 in BPS and TA04-03.MTW give data on wine consumption and heart disease death rates in 19 countries. A scatterplot (Exercise 4.27) shows a moderately strong relationship.

(a) Select **Stat ➤ Basic Statistics ➤ Correlation** from the menu to find the correlation between wine consumption and heart disease deaths. What does a negative correlation say about wine consumption and heart disease deaths? About what percent of the variation among countries in heart disease death rates is explained by the straight-line relationship with wine consumption?

(b) Select **Stat ➤ Regression ➤ Regression** from the menu and click on the Option button to find the least-squares regression line for predicting heart disease death rate from wine consumption and to predict the heart disease death rate in another country where adults average 4 liters of alcohol from wine each year.

(c) The correlation in (a) and the slope of the least-squares line in (b) are both negative. Is it possible for these two quantities to have opposite signs? Explain your answer.

(d) Select **Stat ➤ Regression ➤ Regression** from the menu and click on the Option button to find the predicted heart disease death rate for a country that drinks enough wine to supply 150 liters of alcohol per person. Explain why this result can't be true. Explain why using the regression line for this prediction is not intelligent.

5.26 Table 5.2, TA05-02a.MTW, TA05-02b.MTW, TA05-02c.MTW, and TA05-02d.MTW present four sets of data prepared by the statistician Frank Anscombe to illustrate the dangers of calculating without first plotting the data.

(a) Select **Stat ➤ Basics Statistics ➤ Correlation** to find the correlation and select **Stat ➤ Regression ➤ Regression** to find the least-squares regression line for all four data sets. What do you notice? Use the regression line to predict y for $x = 10$.

(b) Make a scatterplot for each of the data sets and add the regression line to each plot.

(c) In which of the four cases would you be willing to use the regression line to describe the dependence of y on x? Explain your answer in each case.

5.36 Ecologists sometimes find rather strange relationships in our environment. One study seems to show that beavers benefit beetles. The researchers laid out 23 circular plots, each four meters in diameter, in an area where beavers were cutting down cottonwood trees. In each plot, they counted the number of stumps from trees cut by beavers and the number of clusters of beetle larvae. Here and in EX05-36.MTW are the data:

Stumps	2	2	1	3	3	4	3	1	2	5	1	3
Beetle larvae	10	30	12	24	36	40	43	11	27	56	18	40
Stumps	2	1	2	2	1	1	4	1	2	1	4	
Beetle larvae	25	8	21	14	16	6	54	9	13	14	50	

(a) Select **Graph ➤ Scatterplot** to make a scatterplot that shows how the number of beaver-caused stumps influences the number of beetle larvae clusters. What does your plot show? (Ecologists think that the new sprouts from stumps are more tender than other cottonwood growth, so beetles prefer them.)

(b) Select **Stat ➤ Regression ➤ Fitted Line Plot** to find the least-squares regression line with and plot the line through the data.

(c) What percent of the observed variation in beetle larvae counts can be explained by straight-line dependence on stump counts?

5.37 A multimedia statistics learning system includes a test of skill in using the computer's mouse. The software displays a circle at a random location on the computer screen. The subject tries to click in the circle with the mouse as quickly as possible. A new circle appears as soon as the subject clicks the old one. Table 5.3 and TA05-03.MTW give data for one subject's trials, 20 with each hand. Distance is the distance from the cursor location to the center of the new circle, in units whose actual size depends on the size of the screen. Time is the time required to click in the new circle, in milliseconds.

(a) We suspect that time depends on distance. Select **Graph ➤ Scatterplot** and select With Groups to make a scatterplot of time against distance, using separate symbols for each hand.

(b) Describe the pattern. How can you tell that the subject is right-handed?

(c) Select **Graph ➤ Scatterplot** again. This time select With Regression and Groups to make a scatterplot with the regression line of time on distance drawn on your plot separately for each hand. Click on the regression line to find it's equation. Which regression does a better job of predicting time from distance?

5.38 It is possible that the subject in Exercise 5.37 got better in later trials due to learning. It is also possible that he got worse due to fatigue. To examine the data in TA05-03.MTW separately for each hand, select **Data ➤ Unstack Columns** to unstack the time and distance data using the subscripts in the hand column. For each hand, select **Stat ➤ Regression ➤ Regression** and click on the Graphs button to plot the residuals from each regression against the time order of the trials. Is either of these systematic effects of time visible in the data?

5.39 Return to the merlin data regression of Exercise 5.22 and EX05-22.MTW. Select **Stat ➤ Regression ➤ Regression** and click on the Storage button to obtain the residuals. The residuals are the part of the response left over after the straight-line tie to the explanatory variable is removed. Select **Stat ➤ Basic Statistics ➤ Correlation** to find the correlation between the residuals and the explanatory variable. Your result should not be a surprise.

5.40 Select **Stat ➤ Regression ➤ Regression** from the menu and click on the Graphs button to make a residual plot (residual against explanatory variable) for the merlin regression of Exercise 5.22 and EX05-.MTW.

(a) Describe the pattern if we ignore the two years with $x = 38$. Do the $x = 38$ years fit this pattern?

(b) Make an additional copy of the data below the original data. In one set delete the two years with $x = 38$. Add an extra column to the worksheet to mark the complete data and the data with the two deleted years. Select **Graph ➤ Scatterplot** from the menu and select With Regression and Groups to make a scatterplot with two least-squares lines, with all nine years and without the two $x = 38$ years. Although the original regression in Exercise 5.22 seemed satisfactory, the two $x = 38$ years are influential. We would like more data for years with x greater than 33.

5.41 Return to the regression of highway mileage on city mileage in Exercise 5.3 and TA01-02.MTW. Select **Stat ➤ Regression ➤ Regression** and click on the Graphs button to make a residual plot (residuals against city mileage).

(a) Which car has the largest positive residual? The largest negative residual?

(b) The Honda Insight, an extreme outlier, does not have the largest residual in either direction. Why is this not surprising?

(c) Explain briefly what a large positive residual says about a car. What does a large negative residual say?

Chapter 6
Two-Way Tables

Topics to be covered in this chapter:

Tables
Row and Column Percents

Tables

We can describe relationships between two or more categorical variables using two-way or multiway tables in Minitab. We will use the data found in Example 6.5 in BPS and in EG06-05.MTW to show how to make two-way and three-way tables. In this data set, we have stored the information about 1300 accident victims. For each patient, we have information about the patient's outcome (in C1), type of transport (in C2), and the seriousness of the accident (in C3). To determine the number of patients that survived for each type of transportation, select

Stat ➤ Tables ➤ Cross Tabulation and Chi-Square

from the menu. In the dialog box, select the variables to be classified and the type of display. For this example, we choose to display the total number of values (counts) for each cell and for the margins. You can also choose to display row percents, column percents, and total percents.

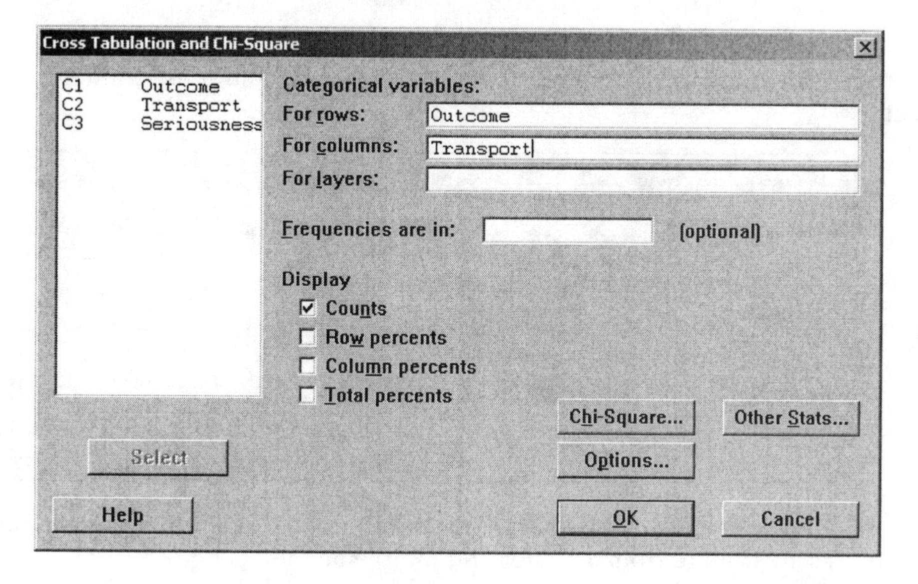

Tabulated statistics: Outcome, Transport

```
Rows: Outcome   Columns: Transport

             Helicopter  Road   All

died                 62   242   304
survived            125   771   896
All                 187  1013  1200

Cell Contents:      Count
```

Note that we selected "Outcome" for the row variable and "Transport" for the column variable. In the dialog box, you can choose more than one type of display at the same time. Check Column percents if you want to display the percentage each cell represents of the total observations in the column. In the following table, the cell contents are expressed as percents of the column values. This two-way table indicates that about 33% of victims transported by helicopter died compared to about 24% of victims transported by road.

Tabulated statistics: Outcome, Transport

```
Rows: Outcome   Columns: Transport

             Helicopter     Road      All

died                 62      242      304
                  33.16    23.89    25.33

survived            125      771      896
                  66.84    76.11    74.67

All                 187     1013     1200
                 100.00   100.00   100.00

Cell Contents:      Count
                    % of Column
```

To make a bar chart comparing the percent of victims that died for the two modes of transportation, enter the appropriate data into the Minitab worksheet.

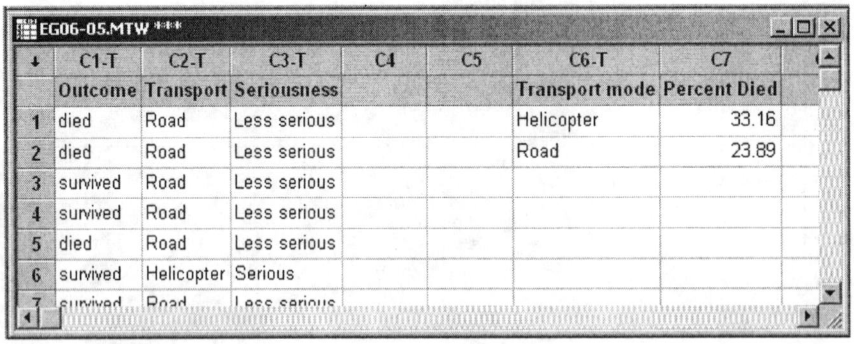

EG06-05.MTW ***

↓	C1-T	C2-T	C3-T	C4	C5	C6-T	C7
	Outcome	Transport	Seriousness			Transport mode	Percent Died
1	died	Road	Less serious			Helicopter	33.16
2	died	Road	Less serious			Road	23.89
3	survived	Road	Less serious				
4	survived	Road	Less serious				
5	died	Road	Less serious				
6	survived	Helicopter	Serious				
7	survived	Road	Less serious				

To make the bar chart, select **Graph ➤ Bar Chart** from the menu. Select Bars represent "Values from a Table" and Simple on the dialog box.

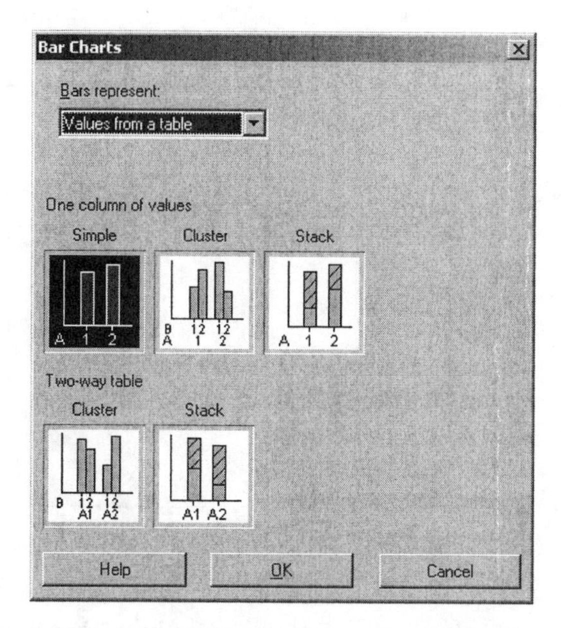

In the next dialog box, indicate the Graph variable "Percent Died" and the Categorical variable "Transport mode" to be graphed.

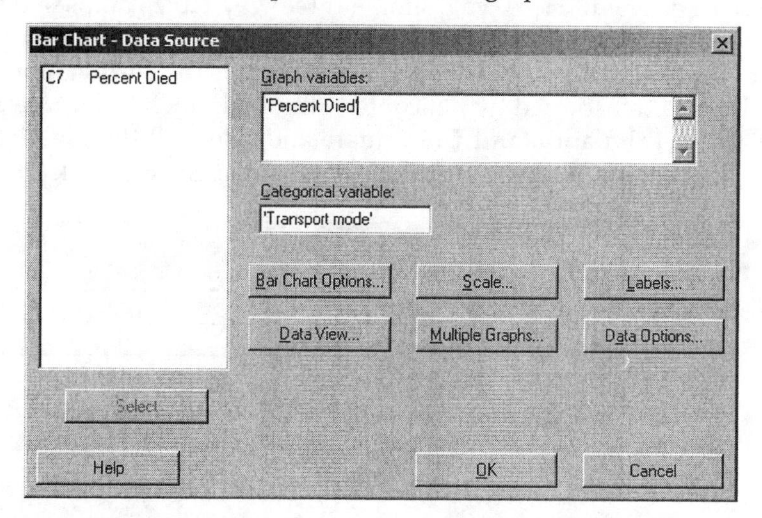

The resulting bar graph shows that when all accidents are combined, victims transported by helicopter are more likely to die. This is because victims of more serious accidents tend to be transported more frequently by helicopter.

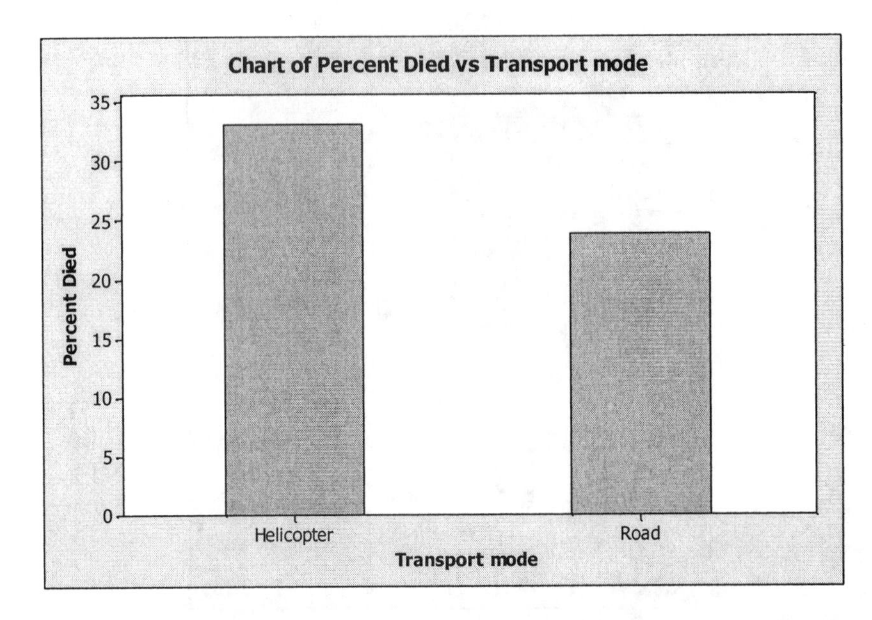

If a third variable is entered For layers in the dialog box, three-way tables will be produced. These are essentially two-way tables for each value of the third variable specified. Here we examine three-way tables and see that for less serious accidents, the survival rate is higher when victims are transported by helicopter. Similarly, we see that for serious accidents the survival rate is higher when victims are transported by helicopter. To obtain these tables, select **Stat ➤ Tables ➤ Cross Tabulation and Chi-Square** and select all three variables to be classified. Choose Column percents as the type of display and click OK.

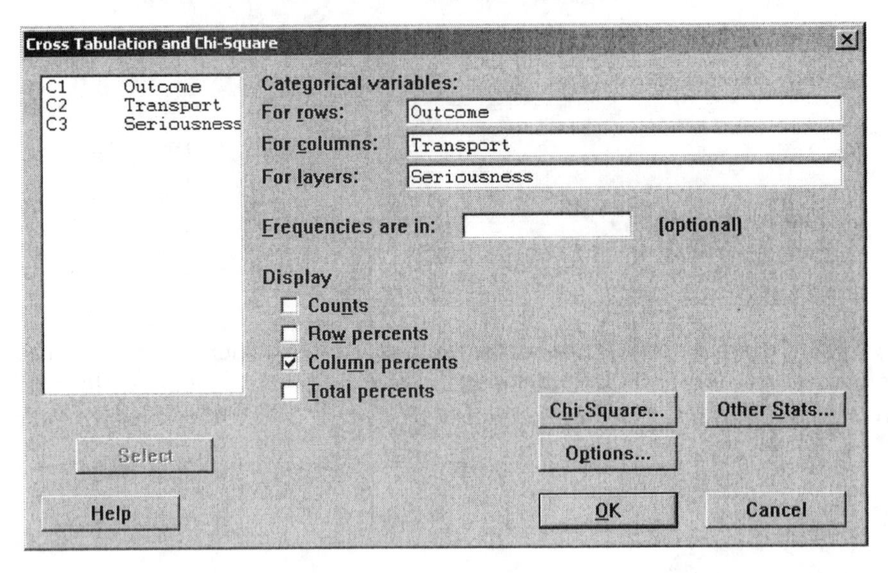

Tabulated statistics: Outcome, Transport, Seriousness

Results for Seriousness = Less serious

```
Rows: Outcome    Columns: Transport

              Helicopter     Road      All

died               17.20    20.09    19.82
survived           82.80    79.91    80.18
All               100.00   100.00   100.00

Cell Contents:        % of Column
```

Results for Seriousness = Serious

```
Rows: Outcome    Columns: Transport

              Helicopter     Road      All

died               48.94    59.79    54.45
survived           51.06    40.21    45.55
All               100.00   100.00   100.00

Cell Contents:        % of Column
```

EXERCISES

6.3 A study in Sweden looked at former elite soccer players, people who had played soccer but not at the elite level, and people of the same age who did not play soccer. Here and in EX06-03.MTW are data that classify these subjects by whether or not they had arthritis of the hip or knee by their mid-50s.

	Elite	Non-elite	Did not play
Arthritis	10	9	24
No arthritis	61	206	548

(a) Select **Stat ➤ Tables ➤ Cross Tabulation and Chi-Square** from the menu to find out how many people these data describe.

(b) How many of these people have arthritis of the hip or knee?

(c) Select **Stat ➤ Tables ➤ Cross Tabulation and Chi-Square** from the menu. Select the appropriate percents to find the marginal distribution of participation in soccer, both as counts and as percents.

(d) Select **Stat ➤ Tables ➤ Cross Tabulation and Chi-Square** from the menu to find the percent of each group in the soccer-risk data that have arthritis. Enter the data into your Minitab worksheet and select **Graph ➤ Bar Chart** to make a bar chart comparing the percents. What do these percents say about the association between playing soccer and later arthritis?

6.7 A study of the career plans of young women and men sent questionnaires to all 722 members of the senior class in the College of Business Administration at the University of Illinois. One question asked which major within the business program the student had chosen. Here and in EX06-07.MTW are the data from the students who responded:

	Female	Male
Accounting	68	56
Administration	91	40
Economics	5	6
Finance	61	59

(a) Select **Stat ➤ Tables ➤ Cross Tabulation and Chi-Square** from the menu to find the two conditional distributions of major, one for women and one for men. Based on your calculations, describe the differences between women and men with a graph and in words.

(b) What percent of the students did not respond to the questionnaire? The nonresponse weakens conclusions drawn from these data.

6.9 Here and in EX06-09.MTW are the numbers of flights on time and delayed for two airlines at five airports in one month. Overall on-time percentages for each airline are often reported in the news. The airport that flights serve is a lurking variable that can make such reports misleading.

	Alaska Airlines		America West	
	On time	Delayed	On time	Delayed
Los Angeles	497	62	694	117
Phoenix	221	12	4840	415
San Diego	212	20	383	65
San Francisco	503	102	320	129
Seattle	1841	305	201	61

(a) What percent of all Alaska Airlines flights were delayed? What percent of all America West flights were delayed? These are the numbers usually reported.

(b) Now find the percent of delayed flights for Alaska Airlines at each of the five airports. Do the same for America West.

(c) America West does worse at every one of the five airports, yet it does better overall. That sounds impossible. Explain carefully, referring to the data, how it can happen. (The weather in Phoenix and Seattle lies behind this example of Simpson's paradox.)

6.10 Whether a convicted murderer gets the death penalty seems to be influenced by the race of the victim. Here and in EX06-10.MTW are data on 326 cases in which the defendant was convicted of murder:

White defendant

	White victim	Black victim
Death	19	0
Not	132	9

Black defendant

	White victim	Black victim
Death	11	6
Not	52	97

(a) Use these data to make a two-way table of defendant's race (white or black) versus death penalty (yes or no).

(b) Show that Simpson's paradox holds: a higher percent of white defendants are sentenced to death overall, but for both black and white victims a higher percent of black defendants are sentenced to death.

(c) Use the data to explain why the paradox holds in language that a judge could understand.

6.16 Here and in EX06-16.MTW are data from eight high schools on smoking among students and among their parents:

	Neither parent smokes	One parent smokes	Both parents smoke
Student does not smoke	1168	1823	1380
Student smokes	188	416	400

(a) Select **Stat ➤ Tables ➤ Cross Tabulation and Chi-Square** from the menu to determine how many students these data describe.

(b) What percent of these students smoke? In the dialog box, should you select the Row percents or Column percents display to answer this question?

(c) Select **Stat ➤ Tables ➤ Cross Tabulation and Chi-Square** from the menu to calculate and compare percents to show how parents' smoking influences students' smoking. Briefly state your conclusions about the relationship.

6.18 Firearms are second to motor vehicles as a cause of nondisease deaths in the United States. Here and in EX06-18.MTW are counts from a study of all firearm-related deaths in Milwaukee, Wisconsin, between 1990 and 1994. We want to compare the types of firearms used in homicides and in suicides. We suspect that long guns (shotguns and rifles) will more often be used in suicides because many people keep them at home for hunting. Make a careful comparison of homicides and suicides, with a bar graph. What do you find about long guns versus handguns?

	Handgun	Shotgun	Rifle	Unknown	Total
Homicides	468	28	15	13	524
Suicides	124	22	24	5	175

6.19 Addicts need cocaine to feel any pleasure, so perhaps giving them an antidepressant drug will help. A 3-year study with 72 chronic cocaine users compared an antidepressant drug called desipramine with lithium and a placebo. (Lithium is a standard drug used to treat cocaine addiction. A placebo is a dummy drug, used so that the effect of being in the study but not taking any drug can be seen.) One-third of the subjects, chosen at random, received each drug. Here and in EX-6-19.MTW are the results:

	Desipramine	Lithium	Placebo
Relapse	10	18	20
No relapse	14	6	4
Total	24	24	24

(a) Compare the effectiveness of the three treatments in preventing relapse. Use percents and draw a bar graph.

(b) Do you think that this study gives good evidence that desipramine actually causes a reduction in relapses?

6.20 Purdue University is a Big Ten university that emphasizes engineering, scientific, and technical fields. University faculty start as assistant professors, then are promoted to associate professor and eventually to professor. Here and in EX-20.MTW is data that break down Purdue's 1621 faculty members in the 1998–1999 academic year by gender and academic rank:

	Female	Male	Total
Assistant professors	126	213	339
Associate professors	149	411	560
Professors	60	662	722
Total	335	1286	1621

(a) Describe the relationship between rank and gender by finding and commenting on several percents.

(b) One possible explanation for the association might be discrimination (women find it harder to win promotion to higher ranks). Suggest other possible explanations.

6.21 People who get angry easily tend to have more heart disease. That's the conclusion of a study that followed a random sample of 12,986 people from three locations for about four years. All subjects were free of heart disease at the beginning of the study. The subjects took the Spielberger Trait Anger Scale, which measures how prone a person is to sudden anger. Here and in EX06-21.MTW are data for the 8474 people in the sample who had normal blood pressure. CHD stands for "coronary heart disease." This includes people who had heart attacks and those who needed medical treatment for heart disease.

	Low anger	Moderate anger	High anger	Total
CHD	53	110	27	190
No CHD	3057	4621	606	8284
Total	3110	4731	633	8474

Present evidence from this two-way table that backs up the study's conclusion about the relationship between anger and heart disease.

6.22 To help consumers make informed decisions about health care, the government releases data about patient outcomes in hospitals. You want to compare Hospital A and Hospital B, which serve your community. Here and in EX06-22.MTW are data on all patients undergoing surgery in a recent time period. The data include the condition of the patient ("good" or "poor") before the surgery. "Survived" means that the patient lived at least 6 weeks following surgery.

Good Condition

	Hospital A	Hospital B
Died	6	8
Survived	594	592
Total	600	600

Serious Condition

	Hospital A	Hospital B
Died	57	8
Survived	1443	192
Total	1500	200

(a) Compare percents to show that Hospital A has a higher survival rate for both groups of patients.

(b) Combine the data into a single two-way table of outcome ("survived" or "died") by hospital (A or B). The local paper reports just these overall survival rates. Which hospital has the higher rate?

(c) Explain from the data, in language a reporter can understand, how Hospital B can do better overall even though Hospital A does better for both groups of patients.

6.23 Wabash Tech has two professional schools, business and law. Here and in EX06-23.MTW are two-way tables of applicants to both schools, categorized by gender and admission decision. (Although these data are made up, similar situations occur in reality.

Business

	Admit	Deny
Male	480	120
Female	180	20

Law

	Admit	Deny
Male	10	90
Female	100	200

(a) Select **Stat ➤ Tables ➤ Cross Tabulation and Chi-Square** from the menu to make a two-way table of gender by admission decision for the combined professional schools by summing entries in the three-way table.

(b) In the dialog box, select the Row percents display to compute separately the percents of male and female applicants admitted from your two-way table. Wabash Tech's professional schools admit a higher percent of male applicants than of female applicants.

(c) Now compute separately the percents of male and female applicants admitted by the business school and by the law school. Select **Stat ➤ Tables ➤ Cross Tabulation and Chi-Square** and select all three variables to be classified. In the dialog box, select the Row percents display and click OK. Each school admits a higher percent of female applicants.

(d) This is Simpson's paradox: both schools admit a higher percent of the women who apply, but overall Wabash admits a lower percent of female applicants than of male applicants. Explain carefully, as if speaking to a skeptical reporter, how it can happen that Wabash appears to favor males when each school individually favors females.

6.25. The EESEE story "Baldness and Heart Attacks" reports results from comparing men under the age of 55 who survived a first heart attack with men admitted to the same hospitals for other reasons. As part of the study, each patient was asked to rate his own degree of baldness, from 1 (no baldness) to 5 (extreme baldness). Here and in EX06-25.MTW are the counts.

Baldness score	Heart attack patients	Other patients
1	251	331
2	165	221
3	195	185
4	50	34
5	2	1

Select **Stat ➤ Tables ➤ Cross Tabulation and Chi-Square** to see if these data show a relationship between degree of baldness and having a heart attack. If they do, what kind of relationship is it? (The EESEE story points out that there are lurking variables, in particular the ages of the men.)

Chapter 7
Producing Data: Sampling

Topics to be covered in this chapter:

Random Samples
Sorting Data

Random Samples

Minitab allows us to select a simple random sample from a population. The sample can be chosen by selecting

Calc ➤ Random Data ➤ Sample From Columns

from the menu. This command randomly samples rows from one or more columns. You can sample with replacement (the same row can be selected more than once), or without replacement (the same row is not selected more than once). If rows are sampled from several columns at once, the same rows are selected from each.

In example 7.6 of BPS, a sample of 5 business clients is to be selected from an accounting firm's list of 30 business clients. The data are stored in EX07-06.MTW. To select the sample, choose **Calc ➤ Random Data ➤ Sample From Columns** from the menu. Specify that you wish to sample 5 rows from column C1 and specify a column to store the result. If you enter a name that is not a current column, a new column will be created. The dialog box follows.

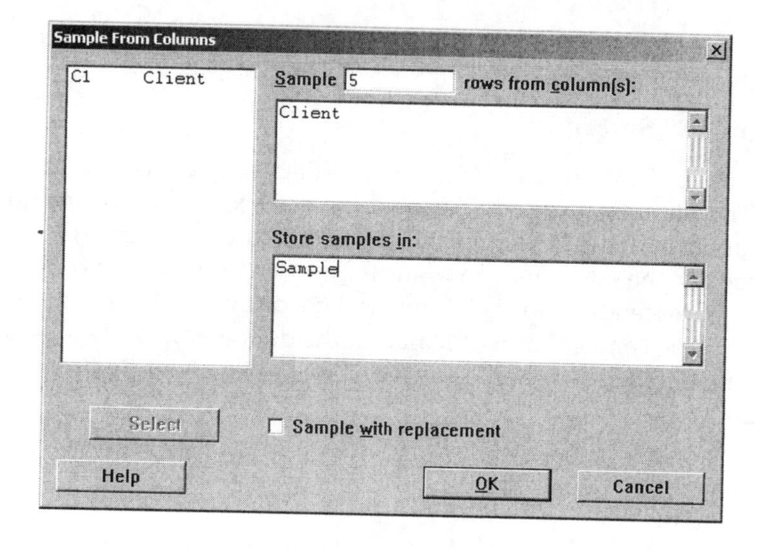

Sometimes it is convenient to use labels instead of names when selecting a random sample. To enter a list of numbers into Minitab, select

Calc ➤ Make Patterned Data ➤ Simple Set of Numbers

from the menu. For example, we can enter the numbers 1 through 1000 into a column of the Minitab worksheet. In the dialog box, specify a column to store the new data, enter the starting point and the end point of the sequence, and click OK.

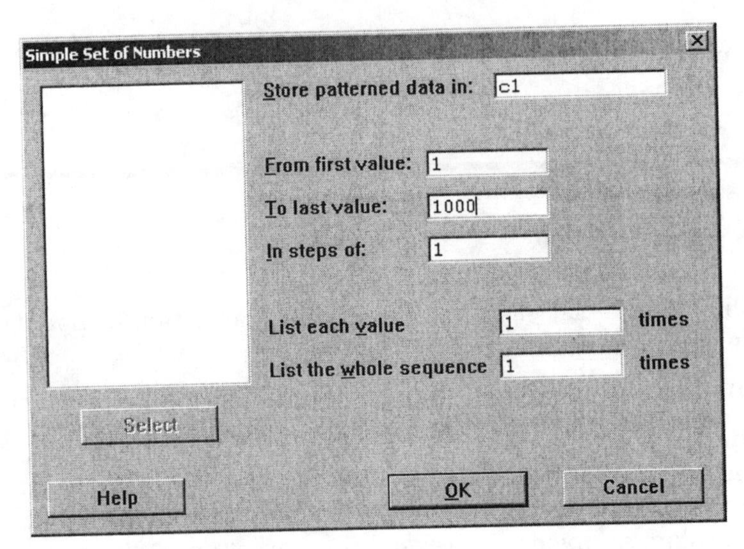

We can then select a sample of individuals to be put into C2 by selecting **Calc ➤ Random Data ➤ Sample From Columns** from the menu. The identification numbers listed in C2 refer to individuals in the population. Once a sample is selected, a population list is needed to determine which individuals are to be included in the sample.

Sorting Data

The samples here were selected in random order. It may be convenient to sort these numbers by selecting

Manip ➤ Sort

from the menu. This command orders the data in a column in numerical sequence. The identification numbers for the individuals to be included in the random sample (C2) can be sorted and put into another column (C3). In the dialog box, specify the columns you want to sort and where you'd like the result stored. If the Descending box is left unchecked, sorting will be done from lowest to highest. The second column specified in the dialog box can be the same as the first column specified. In this case, the sorted data would simply replace the original unsorted data.

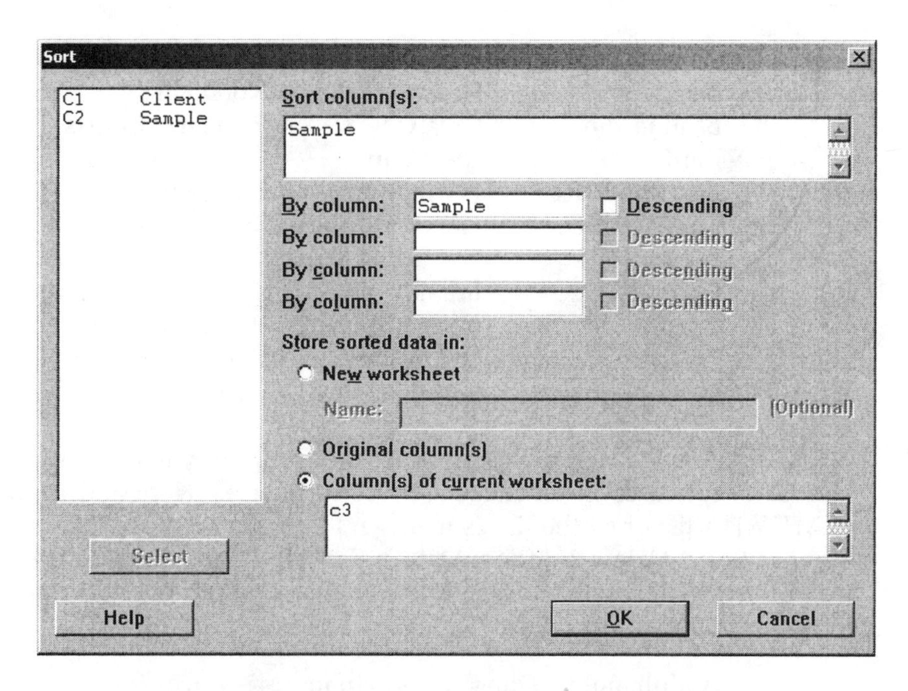

Minitab can sample text as well as numeric data. For text, sorting will alphabetize. To illustrate, we select a sample of five clients to interview from the client list given in Example 7.6 of BPS and sort the sample. First, we enter the client list into a Minitab worksheet. Next, select **Calc ➤ Random Data ➤ Sample From Columns** from the menu to select a sample of five clients from the client list in C1. Finally, select **Manip ➤ Sort** from the menu to list the sample in alphabetical order. The results appear in the following worksheet.

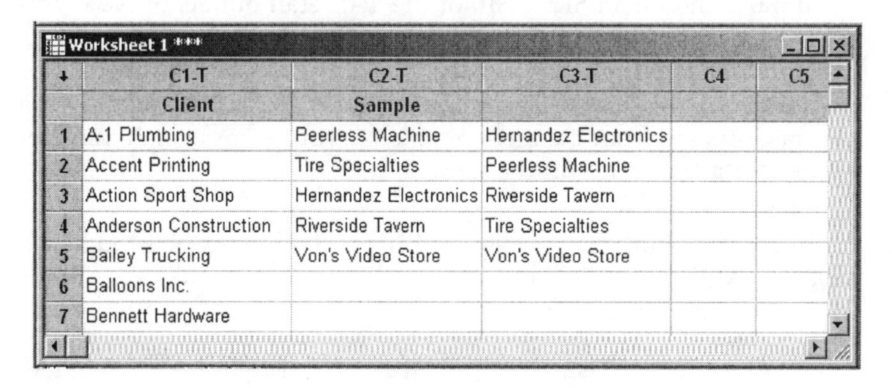

Worksheet 1 ***

↓	C1-T	C2-T	C3-T	C4	C5
	Client	Sample			
1	A-1 Plumbing	Peerless Machine	Hernandez Electronics		
2	Accent Printing	Tire Specialties	Peerless Machine		
3	Action Sport Shop	Hernandez Electronics	Riverside Tavern		
4	Anderson Construction	Riverside Tavern	Tire Specialties		
5	Bailey Trucking	Von's Video Store	Von's Video Store		
6	Balloons Inc.				
7	Bennett Hardware				

EXERCISES

7.8 You are planning a report on apartment living in a college town. You decide to select three apartment complexes at random for in-depth interviews with residents. The list of complexes appears here and in EX07-08.MTW. Select **Calc ➤ Random Data ➤ Sample From Columns** from the menu to select a simple random sample of three of the following apartment complexes.

Ashley Oaks	Country View	Mayfair Village
Bay Pointe	Country Villa	Nobb Hill
Beau Jardin	Crestview	Pemberly Courts
Bluffs	Del-Lynn	Peppermill
Brandon Place	Fairington	Pheasant Run
Briarwood	Fairway Knolls	River Walk
Brownstone	Fowler	Sagamore Ridge
Burberry Place	Franklin Park	Salem Courthouse
Cambridge	Georgetown	Village Square
Chauncey Village	Greenacres	Waterford Court
Country Squire	Lahr House	Williamsburg

7.9 A firm wants to understand the attitudes of its minority managers toward its system for assessing management performance. Here and in EX07-09.MTW is a list of all the firm's managers who are members of minority groups. Select **Calc ➤ Random Data ➤ Sample From Columns** from the menu to choose six to be interviewed in detail about the performance appraisal system.

Abdulhamid	Duncan	Huang	Puri
Agarwal	Fernandez	Kim	Richards
Baxter	Fleming	Liao	Rodriguez
Bonds	Gates	Mourning	Santiago
Brown	Goel	Naber	Shen
Castillo	Gomez	Peters	Vega
Cross	Hernandez	Pliego	Wang

7.10 You must choose an SRS of 10 of the 440 retail outlets in New York that sell your company's products. **Select Calc ➤ Make Patterned Data ➤ Simple Set of Numbers** from the menu to enter the numbers 1 through 440 into a Minitab worksheet. Select **Calc ➤ Random Data ➤ Sample From Columns** from the menu to choose your sample of 10 retail outlets. Select **Data ➤ Sort** to sort the sample.

7.11 A club has 30 student members and 10 faculty members listed here and in EX07-11.MTW.

Students:

Abel	Fisher	Huber	Miranda	Reinmann
Carson	Ghosh	Jimenez	Moskowitz	Santos
Chen	Griswold	Jones	Neyman	Shaw
David	Hein	Kim	O'Brien	Thompson
Deming	Hernandez	Klotz	Pearl	Utts
Elashoff	Holland	Liu	Potter	Varga

Faculty members:

Andrews	Fernandez	Kim	Moore	West
Besicovitch	Gupta	Lightman	Vicario	Yang

The club can send 4 students and 2 faculty members to a convention. It decides to choose those who will go by random selection. Select **Calc ➤ Random Data ➤ Sample From Columns** from the menu to choose a stratified random sample of 4 students and 2 faculty members.

7.12 Accountants use stratified samples during audits to verify a company's records of such things as accounts receivable. The stratification is based on the dollar amount of the item and often includes 100% sampling of the largest items. One company reports 5000 accounts receivable. Of these, 100 are in amounts over $50,000; 500 are in amounts between $1000 and $50,000; and the remaining 4400 are in amounts under $1000. Using these groups as strata, you decide to verify all of the largest accounts and to sample 5% of the midsize accounts and 1% of the small accounts. Select **Calc ➤ Make Patterned Data ➤ Simple Set of Numbers** from the menu to enter the labels 101 to 600 into a column for midsize accounts and the labels 601 to 5000 into another column for small accounts. Select **Calc ➤ Random Data ➤ Sample From Columns** from the menu to select a stratified sample of 25 midsize accounts and 44 small accounts. Select **Data ➤ Sort** to sort the samples.

7.26 You want to ask a sample of college students the question "How much do you trust information about health that you find on the Internet—a great deal, somewhat, not much, or not at all?" You try out this and other questions on a pilot group of 10 students chosen from your class. The class members are here and in EX07-26.MTW.

Anderson	Eckstein	Johnson	Puri
Arroyo	Fernandez	Kim	Richards
Batista	Fullmer	Molina	Rodriguez
Bell	Garcia	Morgan	Samuels
Burke	Gandhi	Nguyen	Shen
Calloway	Glaus	Palmiero	Velasco
Delluci	Helling	Percival	Washburn
Drasin	Husain	Prince	Zhao

Select **Calc ➤ Random Data ➤ Sample From Columns** from the menu to choose your sample. Select **Data ➤ Sort** to sort the sample.

7.32 At a party there are 30 students over age 21 and 20 students under age 21. You choose at random 3 of those over 21 and separately choose at random 2 of those under 21 to interview about attitudes toward alcohol.

(a) Select **Calc ➤ Make Patterned Data ➤ Simple Set of Numbers** from the menu to enter the labels 1 to 30 into a column for students over age 21 and the labels 31 to 50 into another column for students under age 21. Select **Calc ➤ Random Data ➤ Sample From Columns** from the menu to select a stratified sample of 3 students over age 21 and 2 students under age 21.

(b) You have given every student at the party the same chance to be interviewed: What is that chance? Why is your sample not an SRS?

7.34 At a large block party there are 290 men and 110 women. You want to ask opinions about improvements for the next party. To be sure that women's opinions are adequately represented, you decide to choose a stratified random sample of 20 men and 20 women. Select **Calc ➤ Make Patterned Data ➤ Simple Set of Numbers** from the menu to enter the labels 1 to 290 into a column for men and the labels 291 to 400 into another column for women. Select **Calc ➤ Random Data ➤ Sample From Columns** from the menu to select a stratified sample of 20 men and 20 women. Select **Data ➤ Sort** to sort the samples.

Chapter 8
Producing Data: Experiments

Topics to be covered in this chapter:

Randomization in Experiments
Designs
Blocking

Randomization in Experiments

Sampling can be used to randomly select treatment groups in an experiment. In example 8.5 in BPS, 60 family residences will participate in an experiment to reduce energy use. The electric utility company will randomly assign 20 residences to each of the three treatments: Meter, Chart, and Control. By selecting **Calc ➤ Make Patterned Data ➤ Simple Set of Numbers** from the menu, we can enter the numbers 1 through 60 into a Minitab worksheet. To enter the treatment names into column 2, select **Calc ➤ Make Patterned Data ➤ Text values** and fill out the dialog box as illustrated.

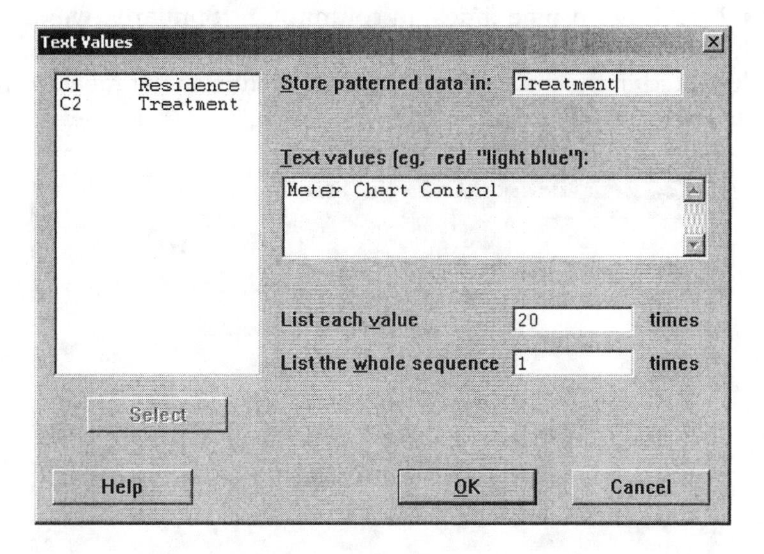

To make the assignment of treatments to residences random, we select **Calc ➤ Random Data ➤ Sample From Columns** from the menu, sample 60 rows from C2, and store the sample back in C2. This will randomize the order of the treatment list in C2. The following worksheet shows the treatments listed in a random order.

↓	C1	C2-T	C3	C4	C5	C6
	Residence	Treatment				
1	1	Control				
2	2	Meter				
3	3	Control				
4	4	Control				
5	5	Meter				
6	6	Chart				
7	7	Meter				

Designs

Sampling can also be used to select treatment groups for more complicated experimental designs. Consider Example 8.2 of BPS. This experiment investigated the effects of repeated exposure to an advertising message. All subjects will view a 40-minute television program that includes ads for a digital camera. Some subjects will see a 30-second commercial, others a 90-second version. The same commercial will be repeated one, three, or five times during the program. This experiment has two factors: length of commercial (with two levels) and repetitions (with three levels). The six different combinations of one level of each factor form six treatments.

First, we create a Minitab worksheet listing the possible treatments and subject identification numbers for 18 subjects. By selecting **Calc ➤ Make Patterned Data ➤ Simple Set of Numbers** from the menu, we can enter the numbers 1 and 2 (each listed nine times) in column C1. Similarly, we can enter the numbers 1, 2, and 3 (each listed three times) in C2 and list the whole sequence twice as illustrated in the dialog box. Finally, we enter the numbers 1 through 18 into C3 to represent the 18 subjects.

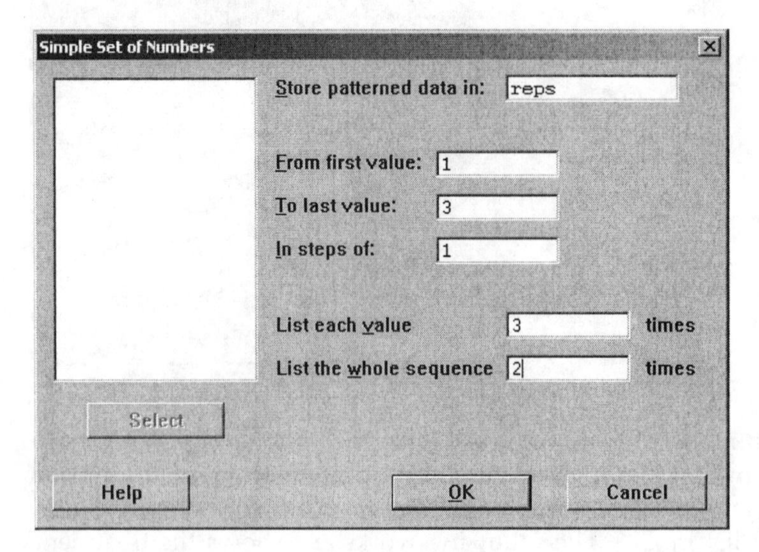

The following worksheet illustrates the experimental design that will be used, that is, three subjects in each of six experimental groups.

↓	C1	C2	C3	C4	C5	C6	C7
	length	reps	subjects				
1	1	1	1				
2	1	1	2				
3	1	1	3				
4	1	2	4				
5	1	2	5				
6	1	2	6				
7	1	3	7				
8	1	3	8				
9	1	3	9				
10	2	1	10				
11	2	1	11				
12	2	1	12				
13	2	2	13				
14	2	2	14				
15	2	2	15				
16	2	3	16				
17	2	3	17				
18	2	3	18				

To randomly assign subject to treatments, select **Calc ➤ Random Data ➤ Sample From Columns** from the menu. Sample 18 rows from C3 and store the sample back in C3. This action will randomize the order of the subjects in C3. The following worksheet shows the subjects listed in a random order so that subjects 6, 12, and 13 are included in the first group (length = 30 seconds, repetitions = 1), subjects 4, 11, and 17 are included in the second group (length = 30 seconds, repetitions = 3), and so on.

↓	C1	C2	C3	C4	C5	C6	C7
	length	reps	subjects				
1	1	1	6				
2	1	1	13				
3	1	1	12				
4	1	2	11				
5	1	2	17				
6	1	2	4				
7	1	3	7				
8	1	3	18				
9	1	3	5				
10	2	1	3				
11	2	1	10				
12	2	1	1				
13	2	2	2				
14	2	2	16				
15	2	2	14				
16	2	3	9				
17	2	3	8				
18	2	3	15				

Blocking

Consider the randomized block design outline in Figure 8.9 of BPS. The blocks consist of male and female subjects, while the treatments are three different television commercials for the same product. The following worksheet contains subject identification numbers for 30 subjects, 1 through 15 in the column for men, 16 through 30 in the column for women. A third column contains the values 1, 2, and 3 corresponding to the different treatments.

↓	C1	C2	C3	C4	C5	C6
	men	women	treatment			
1	1	16	1			
2	2	17	1			
3	3	18	1			
4	4	19	1			
5	5	20	1			
6	6	21	2			
7	7	22	2			
8	8	23	2			
9	9	24	2			
10	10	25	2			
11	11	26	3			
12	12	27	3			
13	13	28	3			
14	14	29	3			
15	15	30	3			
16						

Select **Calc ➤ Random Data ➤ Sample From Columns** to randomize the assignment of subjects. We randomize the order of the men and women separately in their own columns.

↓	C1	C2	C3	C4	C5	C6
	men	women	treatment			
1	4	16	1			
2	10	22	1			
3	2	28	1			
4	6	29	1			
5	8	27	1			
6	14	26	2			
7	12	18	2			
8	15	24	2			
9	5	23	2			
10	9	17	2			
11	13	30	3			
12	11	21	3			
13	1	20	3			
14	3	25	3			
15	7	19	3			
16						

EXERCISES

8.4 The law allows marketers of herbs and other natural substances to make health claims that are not supported by evidence. Brands of ginkgo extract claim to "improve memory and concentration." A randomized comparative experiment found no evidence for such effects. The subjects were 230 healthy people over 60 years old. They were randomly assigned to ginkgo or a placebo pill that looked and tasted the same. All the subjects took a battery of tests for learning and memory before treatment started and again after six weeks.

(a) Outline the design of this experiment. (When you outline the design of an experiment, indicate the size of the treatment groups and the response variable. Figure 8.2 in BPS is a model.)

(b) Select **Calc ➤ Make Patterned Data ➤ Simple Set of Numbers** from the menu to enter numbers 1 to 230 into a Minitab worksheet.

(c) Select **Calc ➤ Random Data ➤ Sample From Columns** to choose the members of the ginkgo group.

8.6 A manufacturer of food products uses package liners that are sealed at the top by applying heated jaws after the package is filled. The customer peels the sealed pieces apart to open the package. What effect does the temperature of the jaws have on the force needed to peel the liner? To answer this question, engineers obtain 20 pairs of pieces of package liner. They seal five pairs at each of 25°F, 27°F, 30°F, and 32°F. Then they measure the force needed to peel each seal.

(a) Use a diagram to describe a completely randomized experimental design for the package liner experiment.

(b) Select **Calc ➤ Make Patterned Data ➤ Simple Set of Numbers** from the menu to enter numbers into a Minitab worksheet corresponding to the experimental units and the treatments. Select **Calc ➤ Random Data ➤ Sample From Columns** to do the randomization required by your design.

8.13 Some investment advisors believe that charts of past trends in the prices of securities can help predict future prices. Most economists disagree. In an experiment to examine the effects of using charts, business students trade (hypothetically) a foreign currency at computer screens. There are 20 student subjects available, named for convenience A, B, C, . . . , T. Their goal is to make as much money as possible, and the best performances are rewarded with small prizes. The student traders have the price history of the foreign currency in dollars in their computers. They may or may not also have software that highlights trends.

(a) Describe two designs for this experiment, a completely randomized design and a matched pairs design in which each student serves as his or her own control.

(b) In both cases, select **Calc ➤ Make Patterned Data ➤ Simple Set of Numbers** from the menu to enter numbers into a Minitab worksheet corresponding to the experimental units and the treatments. Select **Calc ➤ Random Data ➤ Sample From Columns** to do the randomization required by your design and then select **Calc ➤ Random Data ➤ Sample From Columns** to carry out the randomization required by the design.

8.17 The National Institute of Mental Health (NIMH) wants to know whether intense education about the risks of AIDS will help change the behavior of people who report sexual activities that put them at risk of infection. NIMH investigators screened 38,893 people and identified 3706 suitable subjects. The subjects were assigned to a control group (1855 people) or an intervention group (1851 people). The control group attended a one-hour AIDS education session; the intervention group attended seven single-sex discussion sessions, each lasting 90 to 120 minutes. After 12 months, 64% of the intervention group and 52% of the control group said they used condoms. (None of the subjects used condoms regularly before the study began.)

(a) Outline the design of this experiment.

(b) You must randomly assign 3706 subjects. Select **Calc ➤ Patterned Data ➤ Simple Set of Numbers** to enter the numbers 3706 into a Minitab worksheet.

(c) Select **Calc ➤ Random Data ➤ Sample From Columns** to choose the subjects for the intervention group.

8.19 You decide to use a completely randomized design in the two-factor experiment on response to advertising described in Example 8.2 of BPS. The 36 students listed here and in EX08-19.MTW will serve as subjects. Outline the design. Select **Calc ➤ Patterned Data ➤ Simple Set of Numbers** to enter the treatments into a column. Then select **Calc ➤ Random Data ➤ Sample From Columns** to randomly assign the subjects to the 6 treatments.

Alomar	Denman	Han	Liang	Padilla*	Valasco
Asihiro*	Durr*	Howard*	Maldonado	Plochman	Vaughn
Bennett	Edwards*	Hruska	Marsden	Rosen*	Wei
Bikalis	Farouk	Imrani	Montoya*	Solomon	Wilder*
Chao*	Fleming	James	O'Brian	Trujillo	Willis
Clemente	George	Kaplan*	Ogle*	Tullock	Zhang*

8.21 We can improve on the completely randomized design you outlined in Exercise 8.19. The 36 subjects include 24 women and 12 men. The 12 men are marked with asterisks in the list in Exercise 8.19. Men and women often react differently to advertising. You therefore decide to use a block design with the two genders as blocks. You must assign the 6 treatments at random within each block separately.

(a) Outline the design with a diagram.

(b) Select **Data ➤ Unstack Columns** from the menu to create separate columns for the men and women. Select **Calc ➤ Patterned Data ➤ Simple Set of Numbers** to enter the treatments into a column for men and another column for women. Then select **Calc ➤ Random Data ➤ Sample From Columns** to randomly assign the subjects in each block to the 6 treatments. Report your result in a table that lists the 24 women and 12 men and the treatment you assigned to each.

8.29 Does red wine protect moderate drinkers from heart disease better than other alcoholic beverages? Red wine contains substances called polyphenols that may change blood chemistry in a desirable way. This calls for a randomized comparative experiment. The subjects were healthy men aged 35 to 65. They were randomly assigned to drink red wine (9 subjects), drink white wine (9 subjects), drink white wine and also take polyphenols from red wine (6 subjects), take polyphenols alone (9 subjects), or drink vodka and lemonade (6 subjects). Outline the design of the experiment. Select **Calc ➤ Patterned Data ➤ Simple Set of Numbers** to enter subject numbers and treatments into a column. Then select **Calc ➤ Random Data ➤ Sample From Columns** to randomly assign the subjects to the 5 different treatments.

8.34 People who eat lots of fruits and vegetables have lower rates of colon cancer than those who eat little of these foods. Fruits and vegetables are rich in "antioxidants," such as vitamins A, C, and E. Will taking antioxidants help prevent colon cancer? A clinical trial studied this question with 864 people who were at risk of colon cancer. The subjects were divided into four groups: daily beta-carotene, daily vitamins C and E, all three vitamins every day, and daily placebo. After four years, the researchers were surprised to find no significant difference in colon cancer among the groups.

(a) What are the explanatory and response variables in this experiment?

(b) Outline the design of the experiment. Use your judgment in choosing the group sizes.

(c) Select **Calc ➤ Make Patterned Data ➤ Simple Set of Numbers** from the menu to enter numbers into a Minitab worksheet corresponding to the experimental units and the treatments. Select **Calc ➤ Random Data ➤ Sample From Columns** to do the randomization required by your design.

(d) The study was double-blind. What does this mean?

(e) What does "no significant difference" mean in describing the outcome of the study?

(f) Suggest some lurking variables that could explain why people who eat lots of fruits and vegetables have lower rates of colon cancer. The experiment suggests that these variables, rather than the antioxidants, may be responsible for the observed benefits of fruits and vegetables.

8.36 A chemical engineer is designing the production process for a new product. The chemical reaction that produces the product may have higher or lower yield, depending on the temperature and the stirring rate in the vessel in which the reaction takes place. The engineer decides to investigate the effects of all combinations of two temperatures (50°C and 60°C) and three stirring rates (60 rpm, 90 rpm, and 120 rpm) on the yield of the process. She will process two batches of the product at each combination of temperature and stirring rate.

(a) Outline in graphic form the design of an appropriate experiment.

(b) The randomization in this experiment determines the order in which batches of the product will be processed according to each treatment. Select **Calc ➤ Make Patterned Data ➤ Simple Set of Numbers** to enter data for the treatments and experimental units into a Minitab worksheet. Select **Calc ➤ Random Data ➤ Sample From Columns** from the menu to carry out the randomization and state the result.

8.44 Example 8.4 in BPS describes a randomized comparative experiment in which 30 rats are assigned at random to a treatment group of 15 and a control group of 15. Suppose that the 15 even-numbered rats among the 30 rats available are (unknown to the experimenters) a fast-growing variety. We hope that these rats will be roughly equally distributed between the two groups. Select **Calc ➤ Random Data ➤ Sample From Columns** to select a sample of size 15. Record the counts of even-numbered rats in your sample. Repeat for a total of 20 different samples. You see that there is considerable chance variation but no systematic bias in favor of one or the other group in assigning the fast-growing rats. Larger samples from a larger population will on the average do a better job of making the two groups equivalent.

Chapter 9
Introducing Probability

Topics to be covered in this chapter:

Simulating Random Data
Summarizing Results
Simulating Other Distributions

Simulating Random Data

Minitab can be used to simulate random data. To simulate random data, select

Calc ➤ Random Data

from the menu. Then select a distribution from the submenu shown at the right. For example, if you select **Calc ➤ Random Data ➤ Bernoulli** from the menu, you can generate a random sequence of 0s and 1s. In the dialog box, fill in the number of rows to be generated, where they are to be stored, and the probability of success. For example, to simulate a sequence of 50 coin tosses (with equal probability of heads and tails), the dialog box must be filled in as shown on the following page.

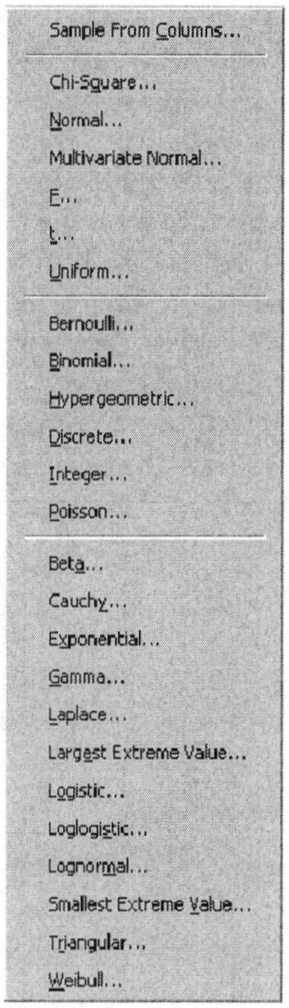

Sample From Columns...
Chi-Square...
Normal...
Multivariate Normal...
F...
t...
Uniform...
Bernoulli...
Binomial...
Hypergeometric...
Discrete...
Integer...
Poisson...
Beta...
Cauchy...
Exponential...
Gamma...
Laplace...
Largest Extreme Value...
Logistic...
Loglogistic...
Lognormal...
Smallest Extreme Value...
Triangular...
Weibull...

For example, the sequence of 0s and 1s may look like:

1, 0, 1, 0, 0, 0, 1, 0, 1, 1, 1, 0, 0, 1, 1, 0, 0, 0, 0, 1, 1, 1, 0, 0, 1, 0, 1, 1, 1, 1, 0, 0, 1, 0, 1, 0, 1, 0, 1, 0, 0, 0, 1, 1, 1, 0, 1, 1, 1, 0.

The 1s correspond to the variable for which you input the probability of success. In this case, we may consider the 1s to be "heads" and the 0s to be "tails." Thus, our sequence of coin flips would be:

H, T, H, T, T, T, H, T, H, H, H, T, T, H, H, T, T, T, T, H, H, H, T, T, H, T, H, H, H, H, T, T, H, T, H, T, H, T, H, T, T, T, H, H, H, T, H, H, H, T.

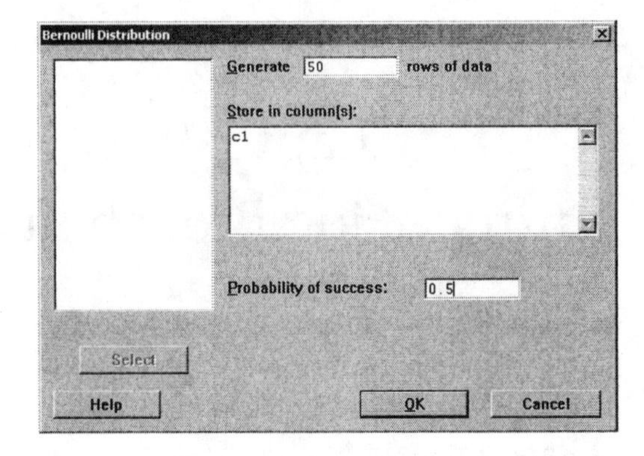

To display the results of the simulation, select

Data ➤ Display Data

from the menu.

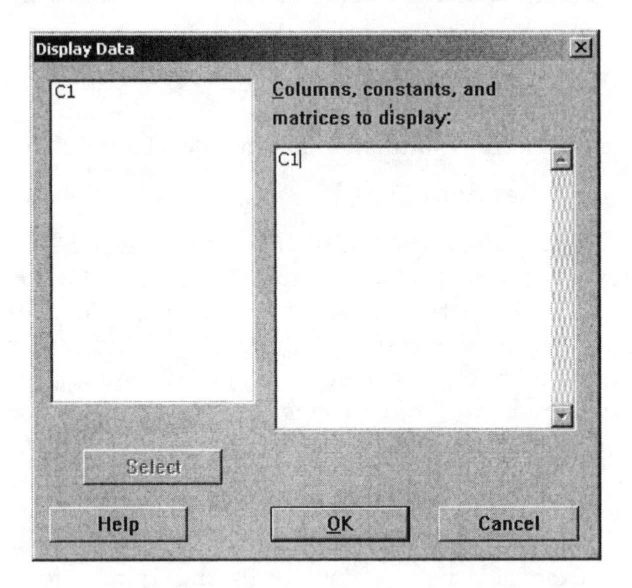

Data Display

```
C1
   1   1   0   0   0   0   0   1   0   0   1   0   1   0   0   1   1   0
   0   1   1   0   0   1   1   0   1   1   1   1   1   1   1   0   0   0
   0   1   0   0   0   1   0   1   1   1   0   0   0   0
```

The command also can be used with several columns at a time to do repeated sampling. To generate 20 replications of the above sample, specify 20 columns, such as C1–C20 in the Bernoulli dialog box.

Summarizing Results

To tabulate results from each column, select

Stat ➤ Tables ➤ Tally

from the Minitab menu. Fill in the variables that you wish to display as follows. Checking the Counts box gives frequency counts for each distinct value in each input column. Checking the Percents box gives percentages for each distinct value in the input column(s), starting at the smallest distinct value. The Cumulative counts box gives cumulative frequency counts for each distinct value in the input column(s), starting at the smallest distinct value and the Cumulative percents box gives cumulative percentage values for each distinct value in the input column(s), starting at the smallest distinct value.

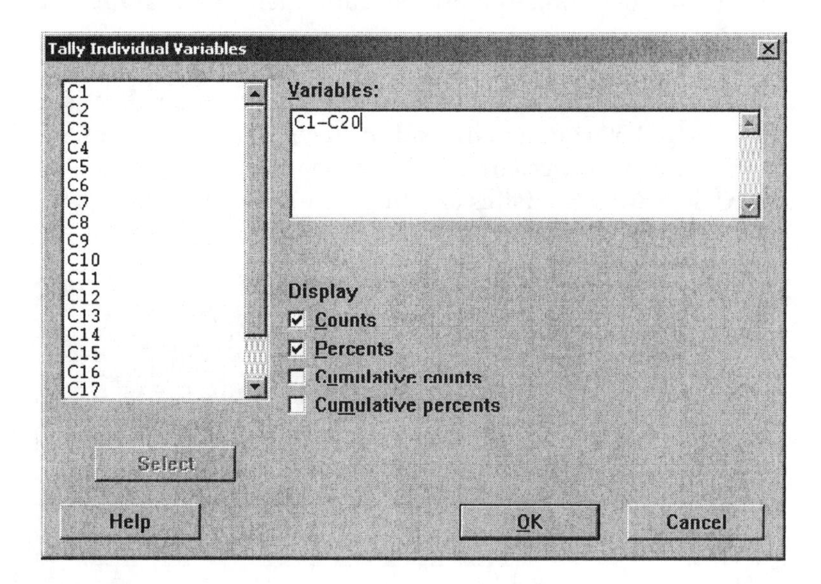

Tally for Discrete Variables: C1, C2, C3, C4, C5, C6, C7, C8, ...

C1	Count	Percent	C2	Count	Percent	C3	Count	Percent
0	31	62.00	0	27	54.00	0	31	62.00
1	19	38.00	1	23	46.00	1	19	38.00
N=	50		N=	50		N=	50	

C4	Count	Percent	C5	Count	Percent	C6	Count	Percent
0	19	38.00	0	18	36.00	0	32	64.00
1	31	62.00	1	32	64.00	1	18	36.00
N=	50		N=	50		N=	50	

C7	Count	Percent	C8	Count	Percent	C9	Count	Percent
0	25	50.00	0	23	46.00	0	24	48.00
1	25	50.00	1	27	54.00	1	26	52.00
N=	50		N=	50		N=	50	

C10	Count	Percent	C11	Count	Percent	C12	Count	Percent
0	29	58.00	0	23	46.00	0	23	46.00
1	21	42.00	1	27	54.00	1	27	54.00
N=	50		N=	50		N=	50	

C13	Count	Percent	C14	Count	Percent	C15	Count	Percent
0	23	46.00	0	22	44.00	0	23	46.00
1	27	54.00	1	28	56.00	1	27	54.00
N=	50		N=	50		N=	50	

C16	Count	Percent	C17	Count	Percent	C18	Count	Percent
0	27	54.00	0	24	48.00	0	22	44.00
1	23	46.00	1	26	52.00	1	28	56.00

```
 N=      50               N=      50               N=      50

C19  Count  Percent   C20  Count  Percent
  0     22    44.00     0     25    50.00
  1     28    56.00     1     25    50.00
 N=     50             N=     50
```

To observe the variability of the simulations, we can display the results of the 20 replications in a histogram and describe the data. To do this, we first need to enter the data into the Minitab worksheet and then select **Graph ➤ Histogram** from the menu.

Another method that will eliminate the need to enter data is to simulate data into 20 rows and 50 columns instead of 50 columns and 20 rows. This way, the row summary may be calculated as row statistics. To calculate the row means, select **Calc ➤ Row Statistics** and fill in the dialog box as shown.

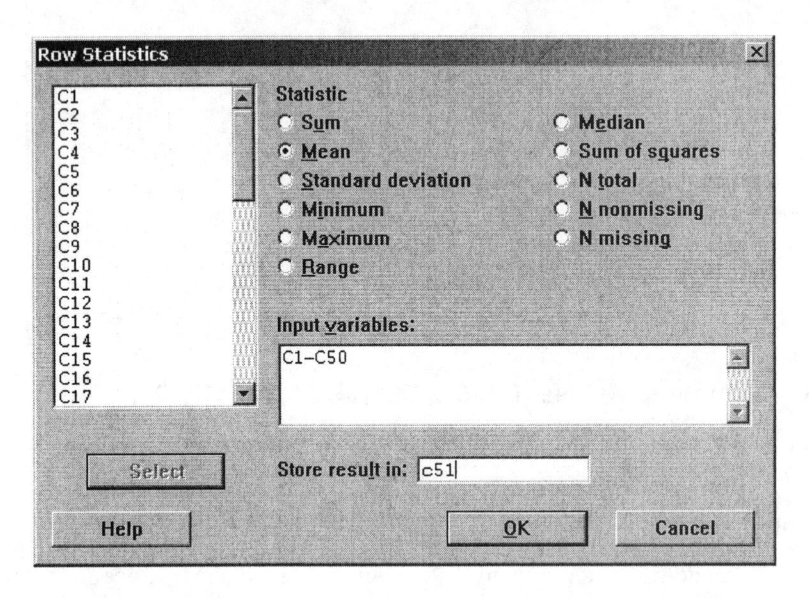

Simulating Other Distributions

In addition to Bernoulli, Minitab can be used to simulate data from many other distributions. For example, discrete distributions can be simulated with Minitab. Example 9.7 in BPS describes Benford's law, a distribution that is often observed in the first digit of numerical records. Here is the distribution for Benford's law.

First digit	1	2	3	4	5	6	7	8	9
Proportion	0.301	0.176	0.125	0.097	0.079	0.067	0.058	0.051	0.046

Any discrete distribution can be specified by putting the values and corresponding probabilities into two columns. We will simulate observations from the distribution following Benford's law. First we enter the sizes and probabilities into a Minitab spreadsheet in the Data window.

↓	C1	C2	C3	C4
	digit	proportion		
1	1	0.301		
2	2	0.176		
3	3	0.125		
4	4	0.097		
5	5	0.079		
6	6	0.067		
7	7	0.058		
8	8	0.051		
9	9	0.046		
10				
11				

To simulate data from a discrete distribution such as the one shown, select **Calc ➤ Random Data ➤ Discrete** from the menu. As shown, you must specify the number of rows you wish to generate, where the data are to be stored, the column specifying the values, and the column specifying the probabilities.

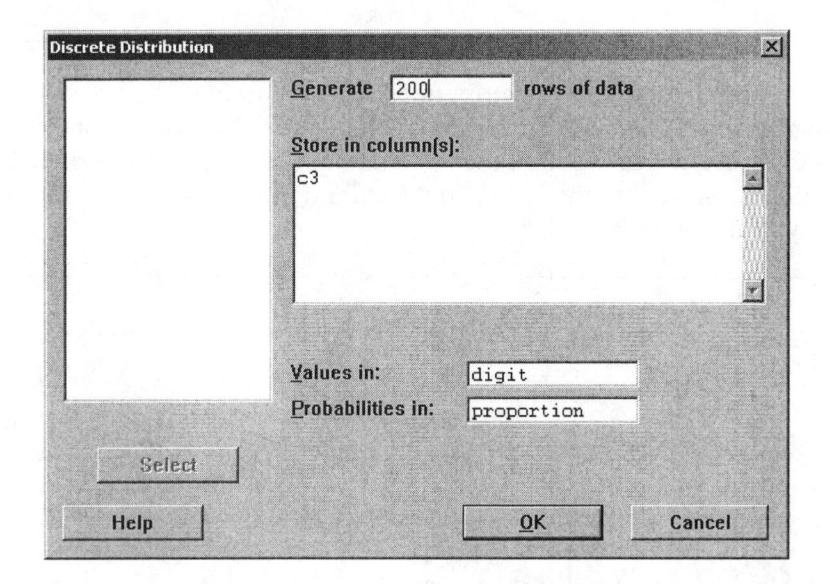

Simulated data will not look exactly like the distribution from which they are selected. To see the difference between the exact distribution and the simulated data, we can compare graphs. To graph the distribution for Benford's law, select **Graph ➤ Bar Chart** from the menu. In the first dialog box, indicate that the bars represent "Values from a table", select "Simple", and click on OK. In the following dialog box, the Graph variable will be C2 or "proportion" and the Categorical variable will be C1 or "digit". Click OK to obtain the following Bar Chart.

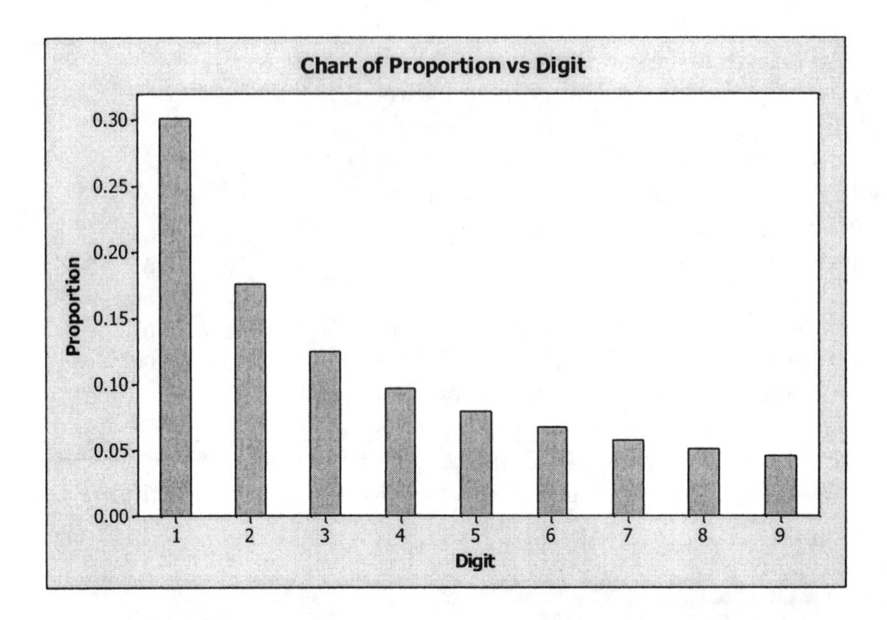

To compare the actual distribution with the simulated data, select **Graph ➤ Histogram** and then select Simple. The Graph variable is C3. Both graphs are skewed to the right, but the simulated data are not as smoothly distributed as the distribution illustrated in the bar chart. The randomness illustrated in the histogram is typical of simulated data.

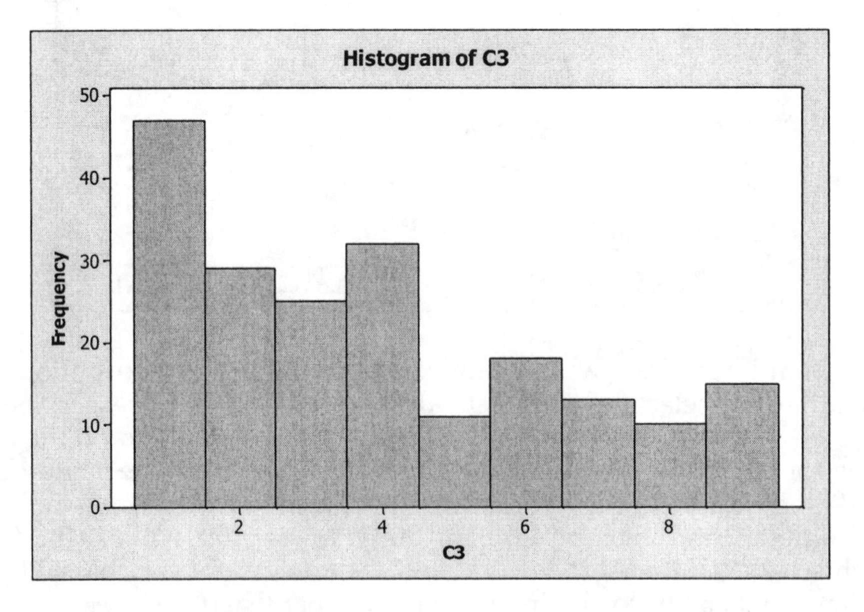

To generate random numbers that are spread out uniformly between two numbers, select **Calc ➤ Random Data ➤ Uniform**. For example to generate 1000 random numbers uniformly across the interval from 0 to 1 described in Example 9.8 of BPS, the dialog box would be filled out as follows.

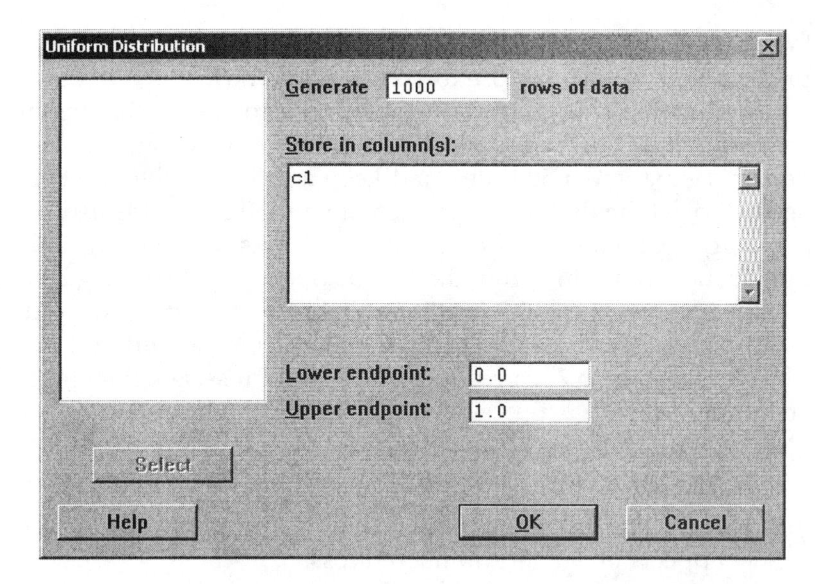

Selecting **Calc ➤ Random Data ➤ Normal** from the menu will simulate observations from a Normal distribution. To simulate the heights of ten young women with the N(64, 2.7) distributions described in Example 9.9, the dialog box would be filled out as follows.

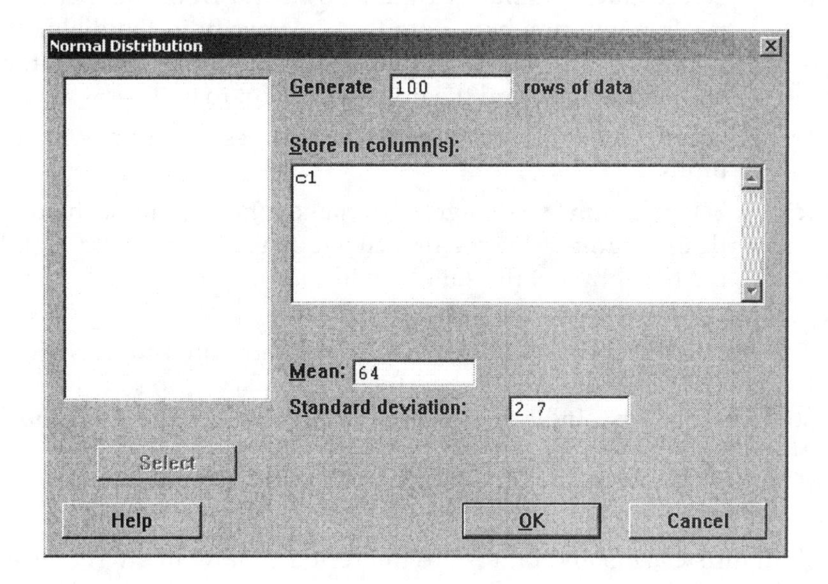

EXERCISES

9.5 Select **Calc ➤ Random Data ➤ Integer** from the menu to simulate random data. Specify in the dialog box that you want 200 integers between 0 and 9. You are using a random mechanism that gives each digit probability 0.1 of being a 0. The table of random digits (Table B) was produced this way. What proportion of the first 200 digits in your list are 0s? This proportion is an estimate, based on 200 repetitions, of the true probability, which in this case is known to be 0.1.

9.6 When we toss a penny, experience shows that the probability (long-term proportion) of heads is close to 1/2. Suppose now that we toss the penny repeatedly until we get heads. What is the probability that the first heads comes up in an odd number of tosses (1, 3, 5, and so on)? To find out, repeat this experiment 50 times, and keep a record of the number of tosses needed to get the first heads on each of your 50 trials. Instead of tossing a penny, generate a sequence of heads and tails by selecting **Calc ➤ Random Data ➤ Bernoulli** from the menu. Specify in the dialog box that you want 300 values using 0.5 for the probability of success. Consider "1" to be heads, so that for example, the sequence starting with 0 1 1 0 0 1 1 0 1, corresponds to 2 tosses, 1 toss, 3 tosses, 1 toss, and 2 tosses needed to get heads.

(a) From your experiment, estimate the probability of heads on the first toss. What value should we expect this probability to have?

(b) Use your results to estimate the probability that the first heads appears on an odd-numbered toss.

9.19 Generate two random numbers between 0 and 1 and take T to be their sum. The sum T can take any value between 0 and 2. The density curve of T is the triangle shown in Figure 4.6 of BPS.

(a) Select **Calc ➤ Random Data ➤ Uniform** from the menu to simulate random numbers from 0 to 1. Specify in the dialog box that you want 1000 values stored in two columns. Specify the lower endpoint to be 0 and the upper endpoint to be 1.

(b) Select **Calc ➤ Calculator** to add the results from your two columns.

(c) Select **Graph ➤ Histogram** to make a histogram of the data in all three columns. Do the first two look similar to a rectangle? Does the histogram of the sum look like a triangle?

9.22 The Normal distribution with mean $\mu = 6.8$ and standard deviation $\sigma = 1.6$ is a good description of the Iowa Test vocabulary scores of seventh-grade students in Gary, Indiana. Select **Calc ➤ Probability Distributions ➤ Normal** from the menu to find the probability that the score of a randomly chosen student is 10 or higher.

9.36 Furniture makers and others are interested in how many rooms housing units have because more rooms can generate more sales. Here are the distributions of the number of rooms for owner-occupied units and renter-occupied units in San Jose, California:

Rooms	1	2	3	4	5	6	7	8	9	10
Owned	.003	.002	.023	.104	.210	.224	.197	.149	.053	.035
Rented	.008	.027	.287	.363	.164	.093	.039	.013	.003	.003

Enter the data into a Minitab worksheet. Select **Graph ➤ Chart** and select Rooms under X and the probability values (Owned and then Rented) under Y to make a probability histogram of these two distributions. What are the most important differences between the distributions for owner-occupied and rented housing units?

9.43 Scores on the National Assessment of Educational Progress twelfth-grade mathematics test for the year 2000 were approximately Normal with mean 300 points (out of 500 possible) and standard deviation 35 points. Let Y stand for the score of a randomly chosen student. Express each of the following events in terms of Y and select **Calc ➤ Probability Distributions ➤ Normal** from the menu to find the probability.

(a) The student's score is above 300.

(b) The student's score is above 370.

9.49 The idea of probability is that the proportion of heads in many tosses of a balanced coin eventually gets close to 0.5. But does the actual count of heads get close to one-half the number of tosses? Let's find out. Instead of tossing a penny, generate a sequence of heads and tails by selecting **Calc ➤ Random Data ➤ Bernoulli** from the menu. Specify in the dialog box that you want 40 values using 0.5 for the probability of success. Select **Stat ➤ Tables ➤ Tally** to find the proportion and count of 1's (corresponding to the number of heads).

(a) For 40 "tosses," what is the proportion of heads? What is the count of heads? What is the difference between the count of heads and 20 (one-half the number of tosses)?

(b) For 120 "tosses," record the proportion and count of heads and the difference between the count and 60 (half the number of tosses).

(c) Keep going. For 240 "tosses" and 480 "tosses" record the same facts. Although it may take a long time, the laws of probability say that the proportion of heads will always get close to 0.5 and also that the difference between the count of heads and half the number of tosses will always grow without limit.

9.50 Over the course of a season, the basketball player Shaquille O'Neal successfully makes about half of his free throws. Select **Calc ➤ Random Data ➤ Bernoulli** from the menu. Specify in the dialog box that you want 100 values using 0.5 for the probability of success to simulate 100 free throws shot independently by a player who has probability 0.5 of making each shot successfully. (Our outcomes here are "Hit" and "Miss.") Select **Stat ➤ Tables ➤ Tally** to find the proportion of 1's (corresponding to the number of Hits).

(a) What percent of the 100 shots did he hit?

(b) Examine the sequence of hits and misses. How long was the longest run of shots made? Of shots missed? (Sequences of random outcomes often show runs longer than our intuition thinks likely.)

9.51 A recent opinion poll showed that about 65% of the American public has a favorable opinion of the software company Microsoft. Suppose that this is exactly true. Then choosing a person at random has probability 0.65 of getting one who has a favorable opinion of Microsoft. Select **Calc ➤ Random Data ➤ Bernoulli** from the menu. Specify in the dialog box that you want 20 values using 0.65 for the probability of success to simulate drawing 20 people. Select **Stat ➤ Tables ➤ Tally** to find the proportion of 1's (corresponding to the number of people with a favorable opinion of Microsoft).

(a) Simulate drawing 20 people, then 80 people, then 320 people. What proportion has a favorable opinion of Microsoft in each case? We expect (but because of chance variation we can't be sure) that the proportion will be closer to 0.65 in longer runs of trials.

(b) Simulate drawing 20 people 10 times by selecting **Calc ➤ Random Data ➤ Binomial** from the menu. Specify in the dialog box that you want 10 values using 20 trials and 0.65 for the probability of success. Record the number in each trial who have a favorable opinion of Microsoft. Then simulate drawing 320 people 10 times and again record the 10 numbers. Which set of 10 results is less variable? We expect the results of 320 trials to be more predictable (less variable) than the results of 20 trials. That is "long-run regularity" showing itself.

Chapter 10
Sampling Distributions

Topics to be covered in this chapter:

The Central Limit Theorem
Control Charts
Out-of-Control Signals

The Central Limit Theorem

Example 10.7 in BPS uses the central limit theorem. The time a technician takes to service an air conditioning unit is exponentially distributed with mean $\mu = 1$ hour and standard deviation $\sigma = 1$ hour. This distribution is strongly right skewed. Figure 10.4 in BPS illustrates the central limit theorem for this distribution. We can also use Minitab to illustrate the central limit theorem.

To generate 250 rows in 25 columns, select **Calc ➤ Random Data ➤ Exponential** from the menu. Specify 250 rows, columns c1-c25 for storage and a mean equal to 1 in the dialog box. To calculate the sampling distribution for the mean of 25 observations, select **Calc ➤ Row Statistics** and fill out the dialog box as follows. Similarly, we can calculate the sampling distribution for the mean of 2 or 5 observations by selecting fewer input variables.

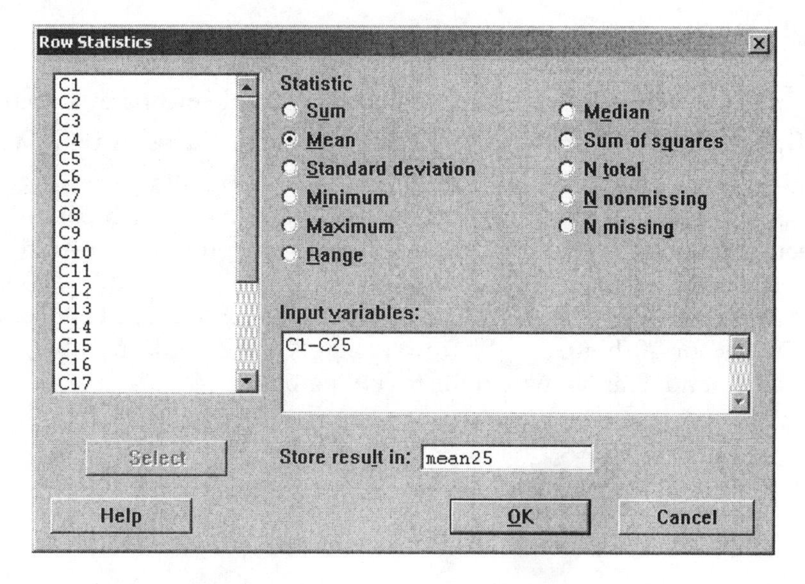

Select **Graph ➤ Histogram** from the menu to produce a histogram of the original data or the mean of 2, 10, or 25 observations. To see how well a Normal distribution fits the data, we select With Fit in the first dialog box. The histograms illustrate the right skewness of the original data and then sample means from 2, 10, and 25 observations. As n increases, the shape of the distribution, becomes more Normal. The mean stays at μ = 1, and the standard deviation decreases.

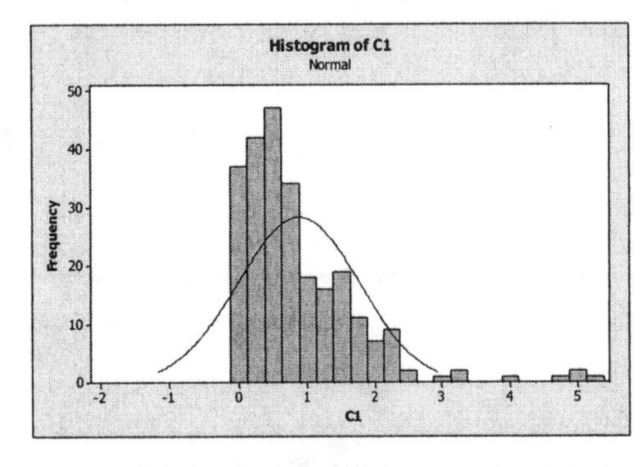

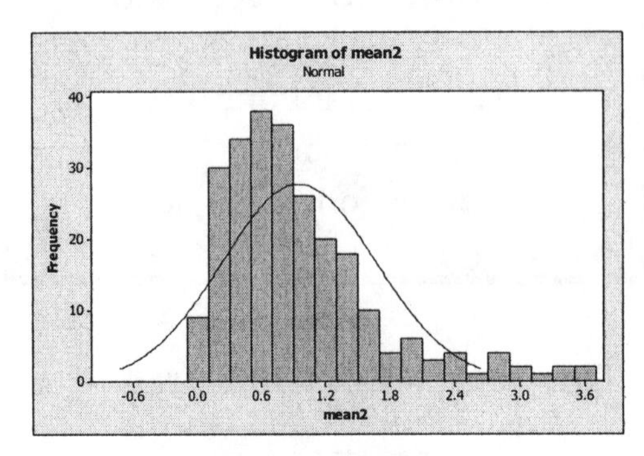

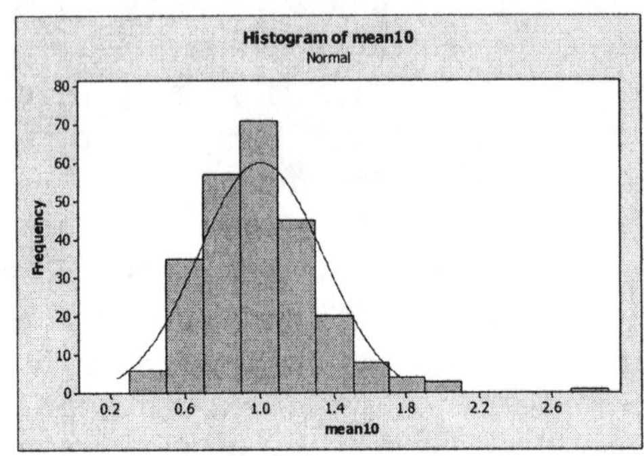

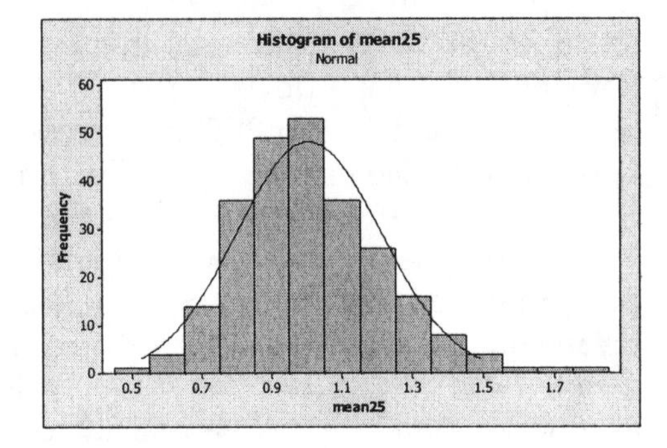

Control Charts

Minitab can be used to produce control charts for sample means by selecting

Stat ➤ Control Charts ➤ Variable Chart for Subgroups ➤ Xbar

from the menu. Example 10.8 of BPS discusses a manufacturer of computer monitors. The manufacturer measures the tension of fine wires behind the viewing screen. The proper tension is 275 mV. When the process is operating properly, the standard deviation of the tension readings is σ = 43 mV. Four measurements are made every hour. Table 10.1 of BPS and TA10-01.MTW contain the measurements for 20 hours. Each row contains the sample number, the four sample values, and $\overline{x}$ as shown on the following page.

TA10-01.MTW ***

↓	C1	C2	C3	C4	C5	C6	C7
	Sample	s1	s2	s3	s4	x-bar	
1	1	234.5	272.3	234.5	272.3	253.4	
2	2	311.1	305.8	238.5	286.2	285.4	
3	3	247.1	205.3	252.6	316.1	255.3	
4	4	215.4	296.8	274.2	256.8	260.8	
5	5	327.9	247.2	283.3	232.6	272.7	
6	6	304.3	236.3	201.8	238.5	245.2	
7	7	268.9	276.2	275.6	240.2	265.2	
8	8	282.1	247.7	259.8	272.8	265.6	

An $\bar{x}$-control chart can be made using either the raw sample data in columns C2–C5 or the $\bar{x}$ data. To make a control chart using the $\bar{x}$ data, specify that the data are arranged as a "Single column" and select the column with the $\bar{x}$ data. You must also specify that the Subgroup size is 1 as shown:

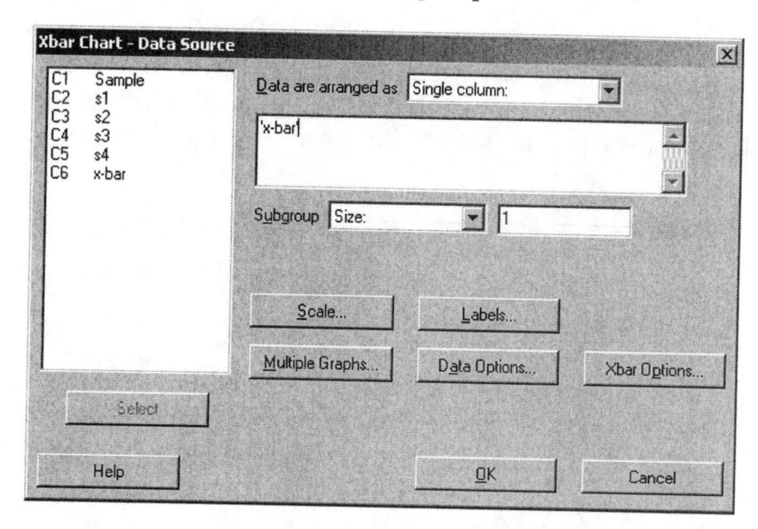

The $\bar{x}$ values will be plotted on the chart. In addition, a center line, an upper control limit (UCL) at 3σ above the center line, and a lower control limit (LCL) at 3σ below the center line are drawn on the chart. By default, the process mean μ and standard deviation σ are estimated from the data. Alternatively, the parameters μ and σ may be specified from historical data by clicking on the Xbar Options button and selecting the Parameters tab. In the dialog box below we specify that the historical mean is equal to 275 and the historical standard deviation is equal to $\sigma/\sqrt{n} = 21.5$.

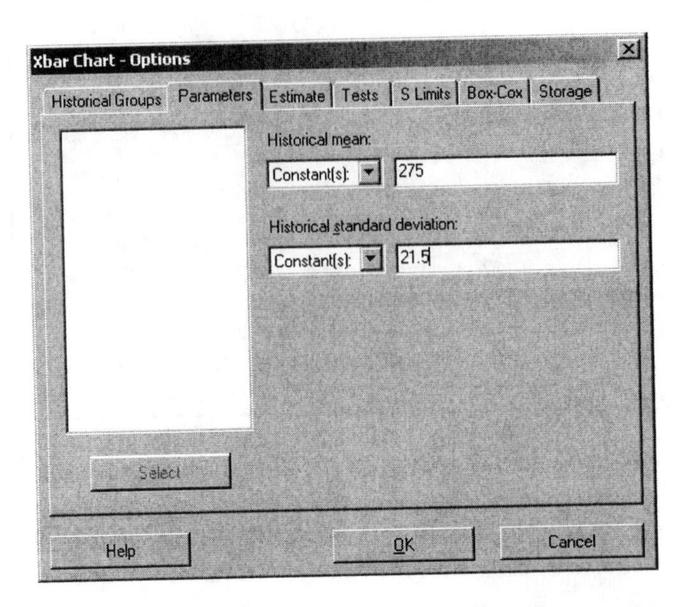

Alternatively, the $\overline{x}$-control chart can be made by specifying that the data are arranged as "subgroups of rows of" and select the columns containing the raw data. In this case, the historical mean is still equal to 275, but the historical standard deviation is equal to $\sigma = 43$. Both methods will produce an $\overline{x}$-control chart as follows.

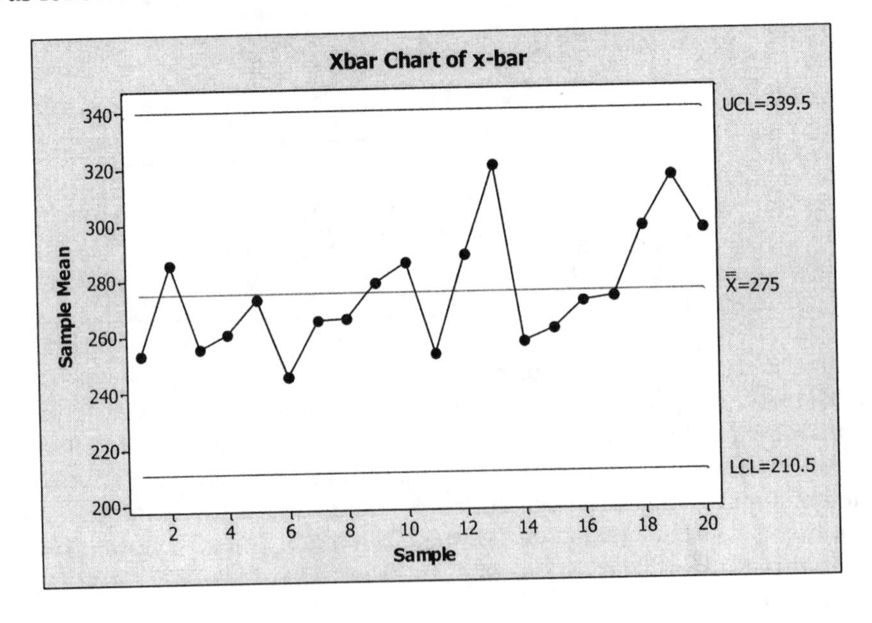

Out-of-Control Signals

Minitab performs tests to identify out-of-control signals. Each test detects a specific pattern in the data plotted on the chart. The occurrence of a pattern suggests a special cause for the variation, one that should be investigated. The tests can be selected by clicking the X-bar Options button on the dialog box and choosing from the Tests tab.

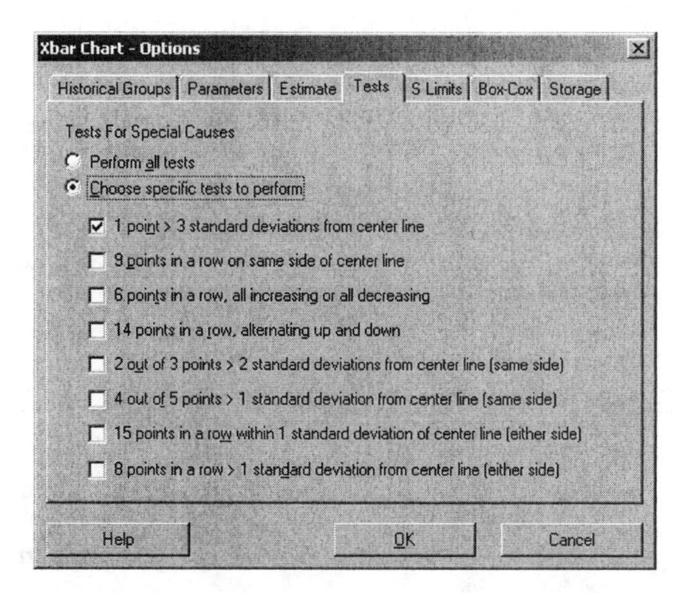

When a point fails a test, it is marked with the test number on the plot. If a point fails more than one test, the number of the first test in your list is the number displayed on the plot.

EXERCISES

10.6 Let us illustrate the idea of a sampling distribution in the case of a very small sample from a very small population. The population is the scores of 10 students on an exam:

Student	1	2	3	4	5	6	7	8	9	10
Size	82	62	80	58	72	73	65	66	74	62

The parameter of interest is the mean size μ in this population. The sample is an SRS of size $n = 4$ drawn from the population.

(a) Enter the data into a Minitab worksheet. Select **Calc ➤ Column Statistics** to find the mean of the 10 sizes in the population. This is the population mean μ.

(b) Select **Calc ➤ Random Data ➤ Sample From Columns** from the menu to draw an SRS of size 4 from this population. Calculate the mean $\bar{x}$ of the sample sizes. This statistic is an estimate of μ.

(c) Repeat this process 10 times. Select **Graph ➤ Histogram** to make a histogram of the 10 values of $\bar{x}$. You are constructing the sampling distribution of $\bar{x}$. Is the center of your histogram close to μ?

10.9 The scores of twelfth-grade students on the National Assessment of Educational Progress year 2000 mathematics test have a distribution that is approximately Normal with mean $\mu = 300$ and standard deviation $\sigma = 35$.

(a) Choose one twelfth-grader at random. What is the probability that his or her score is higher than 300? Higher than 335? Select

Calc ➤ Probability Distributions ➤ Normal from the menu to find the answers.

(b) Now choose an SRS of four 12th-graders. What are the mean and standard deviation of the sample mean score $\bar{x}$ of these four students? Select **Calc ➤ Calculator** to find the answer.

(c) What is the probability that their mean score is higher than 300? Higher than 335? Select **Calc ➤ Probability Distributions ➤ Normal** from the menu to find the answers.

10.11 The scores of students on the ACT college entrance examination in 2001 had mean $\mu = 21.0$ and standard deviation $\sigma = 4.7$. The distribution of scores is only roughly Normal.

(a) What is the approximate probability that a single student randomly chosen from all those taking the test scores 23 or higher? Select **Calc ➤ Probability Distributions ➤ Normal** from the menu to find the answer.

(b) Now take an SRS of 50 students who took the test. Select **Calc ➤ Calculator** to find the mean and standard deviation of the sample mean score $\bar{x}$ of these 50 students. Select **Calc ➤ Probability Distributions ➤ Normal** from the menu to find the approximate probability that the mean score $\bar{x}$ of these students is 23 or higher.

(c) Which of your two Normal probability calculations in (a) and (b) is more accurate? Why?

10.14 A pharmaceutical manufacturer forms tablets by compressing a granular material that contains the active ingredient and various fillers. The hardness of a sample from each lot of tablets is measured in order to control the compression process. The process has been operating in control with mean at the target value $\mu = 11.5$ and estimated standard deviation $\sigma = 0.2$. Table 10.2 and TA10-02.MTW in BPS gives three sets of data, each representing $\bar{x}$ for 20 successive samples of $n = 4$ tablets. One set remains in control at the target value. In a second set, the process mean μ shifts suddenly to a new value. In a third, the process mean drifts gradually.

(a) What are the center line and control limits for an $\bar{x}$-chart for this process?

(b) Select **Stat ➤ Control Charts ➤ Variable Charts for Subgroups ➤ Xbar** from the menu to make an $\bar{x}$-chart for each set of data. Specify that the data are arranged as a "Single column" and that the Subgroup size is equal to one. Enter the target value for the Historical mean and $\sigma/\sqrt{n}$ for the Historical standard deviation. Click OK to draw the control charts. Check that the center line and control limits agree with part (a).

(c) Based on your work in (b) and the appearance of the control charts, which set of data comes from a process that is in control? In which case does the process mean shift suddenly and at about which sample do you think the mean changed? Finally, in which case does the mean drift gradually?

10.24 Shelia's doctor is concerned that she may suffer from gestational diabetes (high blood glucose levels during pregnancy). There is variation both in the actual glucose level and in the blood test that measures the level. A patient is classified as having gestational diabetes if the glucose level is above 140 milligrams per deciliter one hour after a sugary drink is ingested. Shelia's measured glucose level one hour after ingesting the sugary drink varies according to the Normal distribution with $\mu = 125$ mg/dl and $\sigma = 10$ mg/dl.

(a) If a single glucose measurement is made, what is the probability that Shelia is diagnosed as having gestational diabetes? Select **Calc ➤ Probability Distributions ➤ Normal** to find the answer.

(b) If measurements are made instead on four separate days and the mean result is compared with the criterion 140 mg/dl, what is the mean and standard deviation. Select **Calc ➤ Calculator** to find the standard deviation of the sample mean for the four days.

(c) Select **Calc ➤ Probability Distributions ➤ Normal** to find is the probability that Shelia is diagnosed as having gestational diabetes when four measurments are used for the diagnosis.

(d) Find the level L such that there is probability only 0.05 that the mean glucose level of four test results falls above L for Shelia's glucose level distribution. What is the value of L? Select **Calc ➤ Probability Distributions ➤ Normal** to do the required backward Normal calculation.

10.31 Generating a sampling distribution. Exercise 2.23 in BPS and EX02-23.MTW give the survival times of 72 guinea pigs in a medical experiment. Consider these 72 animals to be the population of interest.

(a) Select **Graph ➤ Histogram** to make a histogram of the 72 survival times. This is the population distribution. It is strongly skewed to the right.

(b) Select **Calc ➤ Calculator** to find the mean of the 72 survival times. This is the population mean μ. Mark μ on the x axis of your histogram.

(c) Select **Calc ➤ Random Data ➤ Sample from Columns** to choose an SRS of size $n = 12$. What is the mean survival time $\overline{x}$ for your sample? Mark the value of $\overline{x}$ with a point on the axis of your histogram from (a).

(d) Choose four more SRSs of size 12, using different parts of Table B. Find $\overline{x}$ for each sample and mark the values on the axis of your histogram from (a). Would you be surprised if all five $\overline{x}$'s fell on the same side of μ? Why?

(e) If you chose a large number of SRSs of size 12 from this population and made a histogram of the $\overline{x}$ values, where would you expect the center of this sampling distribution to lie?

10.32 We want to know what percent of American adults approve of legal gambling. This population proportion p is a parameter. To estimate p, take an SRS and find the proportion $\hat{p}$ in the sample who approve of gambling. If we take many SRSs of the same size, the proportion $\hat{p}$ will vary from sample to sample. The distribution of its values in all SRSs is the sampling distribution of this statistic. Figure 10.10 in BPS is a small population. Each circle represents an adult. The colored circles are people who disapprove of legal gambling, and the white circles are people who approve. You can check that 60 of the 100 circles are white, so in this population the proportion who approve of gambling is $p = 60/100 = 0.6$.

(a) Select **Calc ➤ Make Patterned Data ➤ Simple Set of Numbers** to enter the number 0 to 99 into a column. Select **Calc ➤ Random Data ➤ Sample from Columns** to choose an SRS of size 5. What is the proportion $\hat{p}$ of the people in your sample who approve of gambling?

(b) Take 9 more SRSs of size 5. You now have 10 values of the sample proportion $\hat{p}$. What are they? Enter them in one column of a Minitab worksheet.

(c) Select **Graph ➤ Histogram** from the menu to make a histogram of your values of $\hat{p}$. (You have begun to construct the sampling distribution of $\hat{p}$, though of course 10 samples is a small start.)

(d) Taking samples of size 5 from a population of size 100 is not a practical setting, but let's look at your results anyway. How many of your 10 samples estimated the population proportion $p = 0.6$ exactly correctly? Is the true value 0.6 roughly in the center of your sample values?

10.34 A hospital struggling to contain costs investigates procedures on which it loses money. Government standards place medical procedures into diagnostic related groups (DRGs). For example, major joint replacements are DRG 209. The hospital takes from its records a random sample of 8 DRG 209 patients each month. The losses incurred per patient have been in control, with mean \$6400 and standard deviation \$700. Here and in EX10-34.MTW are the mean losses $\bar{x}$ for the samples taken in the next 15 months.

6244	6534	6080	6476	6469	6544	6415	6697
6497	6912	6638	6857	6659	7509	7374	

(a) What are the center line and control limits for an $\bar{x}$ chart for this process?

(b) Select **Stat ➤ Control Charts ➤ Variable Charts for Subgroups ➤ Xbar** from the menu to make an $\bar{x}$-chart for these months. Notice which of the points are out of control. What does the pattern on your chart suggest about the hospital's losses on major joint replacements?

10.41 We can use a discrete distribution to help grasp the idea of a sampling distribution. We will choose an SRS of 10 numbers from a population labeled with the numbers from 1 to 100. That is, in this exercise, the numbers themselves are the population, not just labels for 100 individuals. The mean of the whole numbers 1 to 100 is $\mu = 50.5$. This is the population mean.

(a) Select **Calc ➤ Random Data ➤ Discrete** to choose an SRS of size 10. Which 10 numbers were chosen? What is their mean? This is the sample mean $\bar{x}$.

(b) Although the population and its mean $\mu = 50.5$ remain fixed, the sample mean changes as we take more samples. Take another SRS of size 10. What are the 10 numbers in your sample? What is their mean? This is another value of $\bar{x}$.

(c) Take 8 more SRSs from this same population and record their means. You now have 10 values of the sample mean $\bar{x}$ from 10 SRSs of the same size from the same population. Select **Graph ➤ Histrogram** from the menu to make a histogram of the 10 values and mark the population mean $\mu = 50.5$ on the horizontal axis. Are your 10 sample values roughly centered at the population value μ? (If you kept going forever, your $\bar{x}$ values would form the sampling distribution of the sample mean; the population mean μ would indeed be the center of this distribution.)

10.42 You have a population in which 60% of the individuals approve of legal gambling. You want to take many small samples from this population to observe how the sample proportion who approve of gambling varies from sample to sample. Select **Calc ➤ Random Data ➤ Binomial** from the menu. Specify in the dialog box that you want 50 rows of data.

(a) Specify in the dialog box that you want 50 rows of data. Specify also that the number of trials is equal to 5 and the probability of success is equal to 0.6. Select **Graph ➤ Histogram** to make a histogram of the 50 sample proportions.

(b) Another population contains only 20% who approve of legal gambling. This time, specify in the dialog box that the probability of success is equal to 0.2. Select **Graph ➤ Histogram** to make a histogram of the 50 sample proportions. How do the centers of your two histograms reflect the differing truths about the two populations?

Chapter 11
General Rules of Probability

Topic to be covered in this chapter:

Using Minitab's Calculator for Probability Calculations

Using Minitab's Calculator for Probability Calculations

Minitab's calculator lets you perform mathematical operations and functions. The results of a calculation can be stored in a column or constant. To use the calculator, choose

Calc ➤ Calculator

from the menu. The dialog box shown on the following page will appear. Under Store result in variable, enter a new or existing column or constant. Under Expression, select variables and functions from their respective lists, and click calculator buttons for numbers and arithmetic functions. You can also type the expressions.

Minitab's calculator performs the basic operations of addition (+), subtraction (–), multiplication (*), division (/), and exponentiation (**). All can be used to for calculations based on the probability rules described in BPS.

Example 11.2 in BPS describes British bomber missions where the probability of losing the bomber was 0.05. The probability that the bomber returned safely from a mission was therefore 0.95. It is reasonable to assume that missions are independent. The probability of surviving 20 missions is

$$P(A_1 \text{ and } A_2 \text{ and } \ldots \text{ and } A_{20}) = P(A_1)P(A_2)\cdots P(A_{20}) = (0.95)^{20}$$

To calculate this probability, select **Calc ➤ Calculator** from the menu, enter the expression to be calculated '(.95)**20)', a place for storing the result C1, and click OK. The result of the calculation appears in the worksheet in C1. To print the result in the session window, select Data ➤ Display Data from the menu, enter C1 in the dialog box, and click on OK. The result follows.

Data Display

```
C1
   0.358486
```

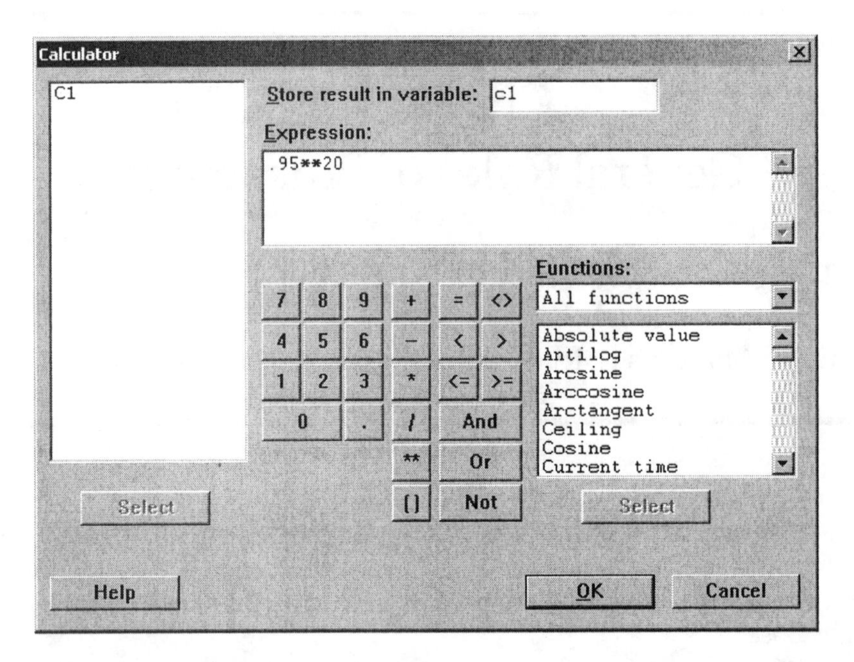

The calculator can be used for any of the expressions involving conditional probabilities and addition and multiplication rules. Example 11.5 in BPS describes a situation in which Deborah and Matthew are anxiously awaiting word on whether they have been made partners of their law firm. Deborah guesses that her probability of making partner is 0.7 and that Matthew's is 0.5. If Deborah also guesses that the probability that *both* she and Matthew are made partners is 0.3, then by the general addition rule

$$P(\text{at least one is promoted}) = 0.7 + 0.5 - 0.3$$

Select **Calc ➤ Calculator** from the menu, enter the appropriate expression, and click OK to calculate this probability.

EXERCISES

11.4 Telephone marketers and opinion polls use random-digit-dialing equipment to call residential telephone numbers at random. The telephone polling firm Zogby International reports that the probability that a call reaches a live person is 0.2. Calls are independent. Select **Calc ➤ Calculator** from the menu to answer the following questions.

(a) A telemarketer places five calls. What is the probability that none of them reaches a person?

(b) When calls are made to New York City, the probability of reaching a person is only 0.08. What is the probability that none of five calls made to New York City reaches a person?

11.5 A string of holiday lights contains 20 lights. The lights are wired in series, so that if any light fails the whole string goes dark. Each light has probability 0.02 of failing during the holiday season. The lights fail independently of each other. Select **Calc ➤ Calculator** from the menu to find the probability that the string of lights will remain bright.

11.6 Slot machines are now video games, with winning determined by electronic random number generators. Suppose you are playing with an old slot machine. You pull the lever to spin three wheels; each wheel has 20 symbols, all equally likely to show when the wheel stops spinning; the three wheels are independent of each other. Suppose that the middle wheel has 9 bells among its 20 symbols, and the left and right wheels have 1 bell each.

(a) You win the jackpot if all three wheels show bells. Select **Calc ➤ Calculator** from the menu to find the probability of winning the jackpot.

(b) There are three ways that the three wheels can stop showing two bells and one symbol other than a bell. Select **Calc ➤ Calculator** from the menu to find the probability of each of these ways. What is the probability that the wheels stop with exactly two bells showing among them?

11.7 Musical styles other than rock and pop are becoming more popular. A survey of college students finds that 40% like country music, 30% like gospel music, and 10% like both. Make a Venn diagram showing these results. Select **Calc ➤ Calculator** from the menu to answer the following questions.

(a) What percent of college students like country but not gospel?

(b) What percent like neither country nor gospel?

11.9 Here are the counts (in thousands) of earned degrees in the United States in the 2001–2002 academic year, classified by level and by the sex of the degree recipient. Select **Calc ➤ Calculator** from the menu to answer the following questions.

	Bachelor's	Master's	Professional	Doctorate	Total
Female	645	227	32	18	922
Male	505	161	40	26	732
Total	1150	388	72	44	1654

(a) If you choose a degree recipient at random, what is the probability that the person you choose is a woman?

(b) What is the conditional probability that you choose a woman, given that the person chosen received a professional degree?

11.19 Chose a student in grades 9 to 12 at random and ask if he or she is studying a language other than English. Here is the distribution of results:

Language	Spanish	French	German	All others	None
Probability	0.26	0.09	0.03	0.03	0.59

What is the conditional probability that a student is studying Spanish, given that he or she is studying some language other than English? Select **Calc ➤ Calculator** from the menu to find the answer.

11.21 An automobile manufacturer buys computer chips from a supplier. The supplier sends a shipment of which 5% fail to conform to performance specifications. Each chip chosen from this shipment has probability 0.05 of being nonconforming, and each automobile uses 12 chips selected independently. What is the probability that all 12 chips in a car will work properly? Select **Calc ➤ Calculator** from the menu to find the answer.

11.22 The "random walk" theory of securities prices holds that price movements in disjoint time periods are independent of each other. Suppose that we record only whether the price is up or down each year, and that the probability that our portfolio rises in price in any one year is 0.65. (This probability is approximately correct for a portfolio containing equal dollar amounts of all common stocks listed on the New York Stock Exchange.) Select **Calc ➤ Calculator** from the menu to answer the following questions.

(a) What is the probability that our portfolio goes up for three consecutive years?

(b) If you know that the portfolio has risen in price two years in a row, what probability do you assign to the event that it will go down next year?

(c) What is the probability that the portfolio's value moves in the same direction in both of the next two years?

11.23 Exercise 11.9 gives the counts (in thousands) of earned degrees in the United States in 2001–2002. Select **Calc ➤ Calculator** from the menu and use these data to answer the following questions.

(a) What is the probability that a randomly chosen degree recipient is a man?

(b) What is the conditional probability that the person chosen received a bachelor's degree, given that he is a man?

(c) Use the multiplication rule to find the probability of choosing a male bachelor's degree recipient. Check your result by finding this probability directly from the table of counts.

11.28 Call a household prosperous if its income exceeds $100,000. Select an American household at random, and let A be the event that the selected household is prosperous and B the event that it is educated. According to the Current Population Survey, $P(\text{A}) = 0.134$, $P(\text{B}) = 0.254$, and the probability that a household is both prosperous and educated is $P(\text{A and B}) = 0.080$. Select **Calc ➤ Calculator** from the menu and use these data to answer the following questions.

(a) Find the conditional probability that a household is educated, given that it is prosperous.

(b) Find the conditional probability that a household is prosperous, given that it is educated.

(c) Are events A and B independent? How do you know?

11.36 A company that offers courses to prepare would-be MBA students for the GMAT examination finds that 40% of its customers are currently undergraduate students and 60% are college graduates. After completing the course, 50% of the undergraduates and 70% of the graduates achieve scores of at least 600 on the GMAT. Select **Calc ➤ Calculator** from the menu to find the percent of all customers who score at least 600 on the GMAT. Select **Calc ➤ Calculator** from the menu and use the following data below to answer the questions in Exercises 11.38 to 11.40.

Highest education	Total population	In labor force	Employed
Did not finish high school	27,325	12,073	11,139
High school but no college	57,221	36,855	35,137
Less than bachelor's degree	45,471	33,331	31,975
College graduate	47,371	37,281	36,259

11.38 Find the unemployment rate for people with each level of education. (This is the conditional probability of being unemployed, given an education level.) How does the unemployment rate change with education? Explain carefully why your results show that level of education and being employed are not independent.

11.39 (a) What is the probability that a randomly chosen person 25 years of age or older is in the labor force?

(b) If you know that the person chosen is a college graduate, what is the conditional probability that he or she is in the labor force?

(c) Are the events "in the labor force" and "college graduate" independent? How do you know?

11.40 You know that a person is employed. What is the conditional probability that he or she is a college graduate? You know that a second person is a college graduate. What is the conditional probability that he or she is employed?

Chapter 12
Binomial Distributions

Topics to be covered in this chapter:

Binomial Probabilities
Normal Approximation to the Binomial

Binomial Probabilities

In Example 12.5 of BPS, an engineer chooses a sample of ten switches from a shipment. Suppose that 10% of the switches in the shipment are bad. The engineer will count the number *X* of bad switches. In Chapter 10 we learned to generate random numbers for this situation by selecting **Calc ➤ Random Data ➤ Bernoulli** from the menu. We can generate a sequence of ten 1's and 0's to represent the bad and good switches:

Data Display

```
C1
   0   0   0   1   0   0   0   0   0   0
```

If we are interested only in the number of bad switches, we can generate the number *X* by selecting **Calc ➤ Random Data ➤ Binomial** from the menu. Instead of generating only one value for *X*, we can also select **Calc ➤ Random Data ➤ Binomial** to simulate a large number of repetitions of the sample.

In addition to simulating binomial data, we can use Minitab to calculate exact binomial probabilities. To use Minitab to calculate a probability, select

Calc ➤ Probability Distributions ➤ Binomial

from the menu. To calculate the probability that one switch is bad (for Example 12.5 in BPS), the dialog box should be filled in with an Input constant of 1. You must also check Probability and specify that the Number of trials is 10 and the Probability of success is 0.1 in the dialog box.

If you want the entire probability distribution, enter the numbers 0 through 10 in a column on the Minitab worksheet. Then select that column as the Input column. You may also select another column for Optional storage.

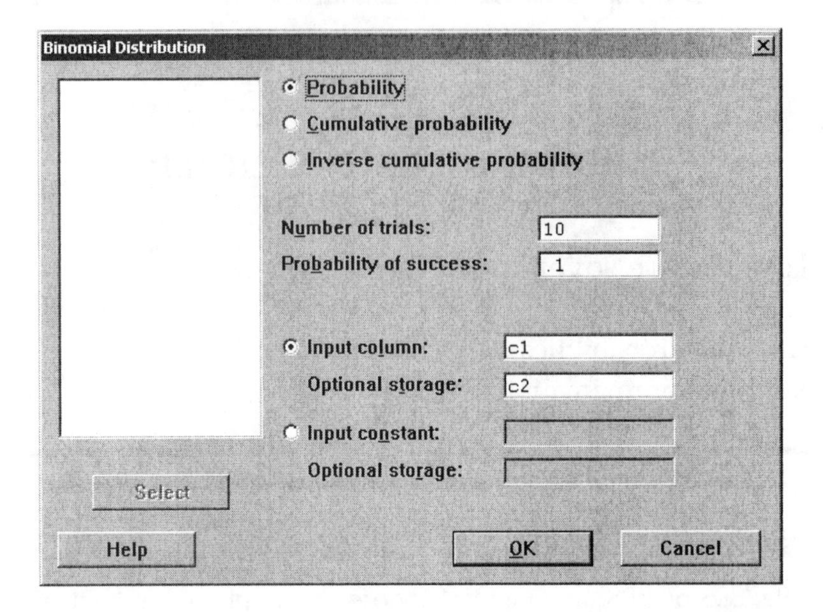

If no column is selected for optional storage, then the results will appear in the session window. Here we obtain the probability of each possible outcome for a binomial distribution with $n = 10$ and $p = 0.1$.

Probability Density Function

```
Binomial with n = 10 and p = 0.1

 x  P( X = x )
 0    0.348678
 1    0.387420
 2    0.193710
 3    0.057396
 4    0.011160
 5    0.001488
 6    0.000138
 7    0.000009
 8    0.000000
 9    0.000000
10    0.000000
```

Note that for $x = 7$, $P(X = x)$ is equal to 0 (rounded to four decimal places). For K = 8, 9, and 10, $P(X = x)$ is also equal to 0, so these rows are not printed in the table of probabilities for the binomial.

To graph the distribution, select **Graph ➤ Bar Charts** from the menu. Select Simple and Values from a Table in the first dialog box. In the following dialog box, enter the probabilities for the Graph variable and the x values for the Categorical variable, and click OK.

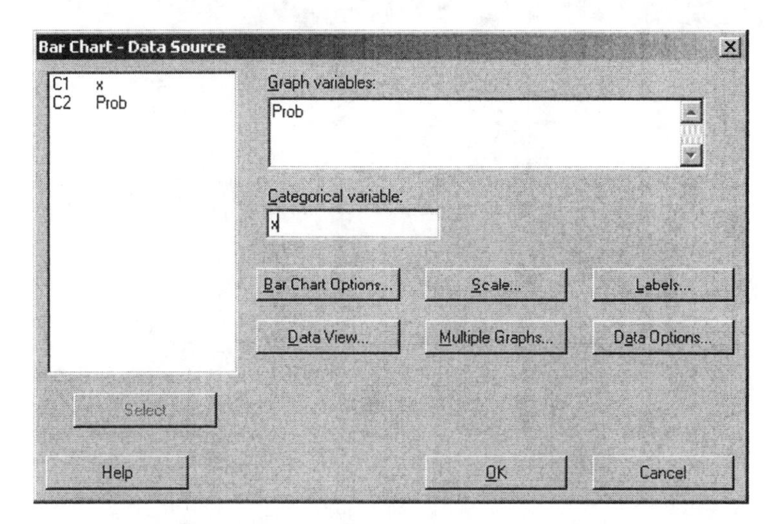

If you select Cumulative Probability in the binomial probability dialog box, then Minitab calculates $P(X \leq x)$ instead of $P(X = x)$. If you wish to calculate $P(X \geq x)$, then it is necessary to realize that $P(X \geq x) = 1 - P(X \leq x{-}1)$.

Suppose that an opinion poll (as described in BPS Example 12.7) asks 2500 adults whether they agree or disagree that "I like buying new clothes, but shopping is often frustrating and time-consuming." Suppose also that 60% of all adult U.S. residents would say "Agree." To find $P(X \geq 1520)$, the probability that at least 1520 adults agree, select **Calc ➤ Probability Distributions ➤ Binomial** from the menu and select Cumulative Probability for 2500 trials and 0.6 Probability of success. As shown, the result is 0.7869. Therefore $P(X \geq 1520) \approx 0.21$.

Cumulative Distribution Function

```
Binomial with n = 2500 and p = 0.6

   x  P( X <= x )
1519     0.786861
```

Normal Approximation to the Binomial

To illustrate the shape of the distribution on the number of adults that would say "Agree" out of the 2500 adults polled, select **Calc ➤ Make Patterned Data ➤ Simple Set of Numbers** from the menu. In the dialog box, specify that you want the numbers from 1350 to 1650 in steps of 1 to be stored in a column of your choice. Next select **Calc ➤ Probability Distributions ➤ Binomial** from the menu and specify that you want the probabilities calculated for 2500 trials with 0.6 probability of success. Use the input column with the numbers from 1350 to 1650 and select another column to store the probabilities. Finally, select **Graph ➤ Scatterplot** from the menu and click on Simple to illustrate the shape. The numbers 1350 to 1650 are the X values and the probabilities are the Y values as shown in the following dialog box.

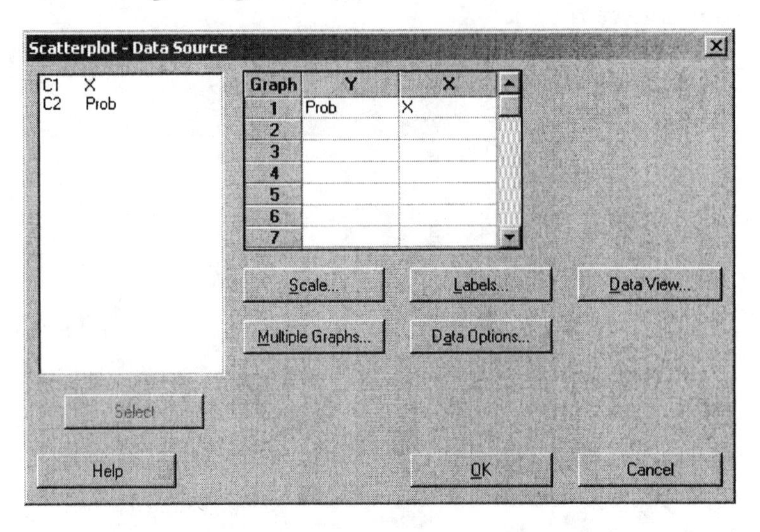

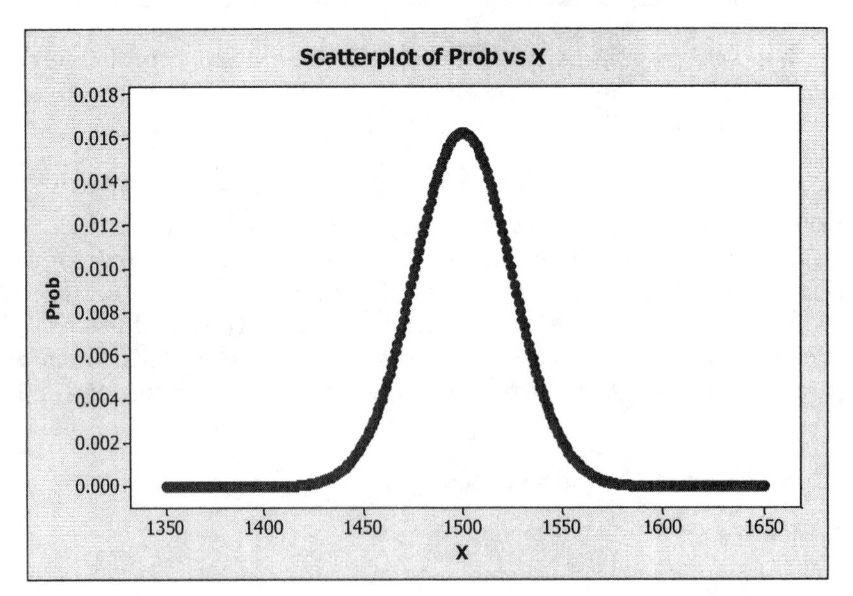

As the figure shows, the binomial probabilities will be approximated well by a normal distribution. The values for the mean and standard deviation are equal to

$$\mu = np = 2500 \times 0.6 = 1500$$

$$\sigma = \sqrt{np(1-p)} = \sqrt{2500 \times 0.6 \times 0.4} = 24.4949.$$

The values are easily calculated using Minitab's calculator. Choose **Calc ➤ Calculator** from the menu and enter "2500*.6" for the mean or "sqrt(250*.6*.4)" for the standard deviation.

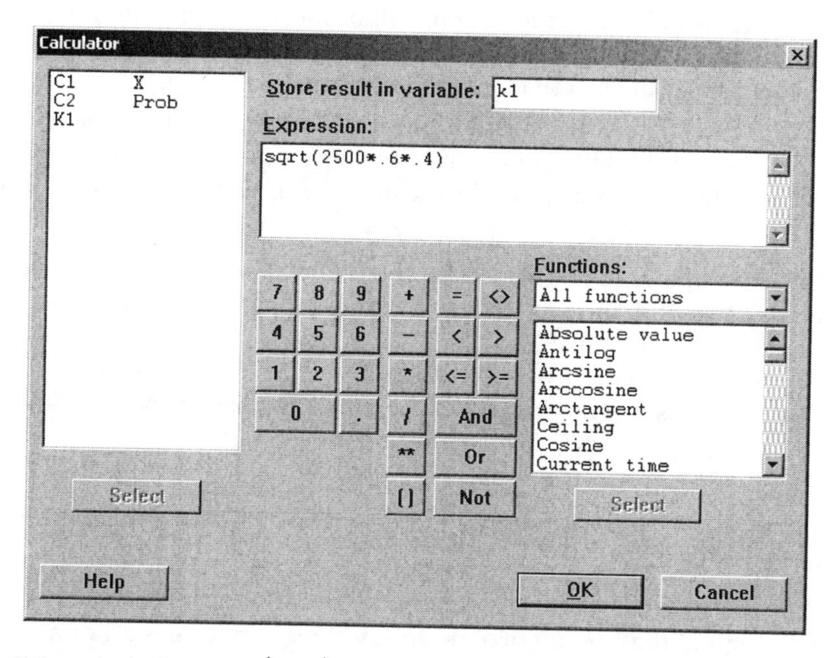

When $np \geq 10$ and $n(1-p) \geq 10$, we can use the normal approximation to approximate binomial probabilities. Here, we approximate the probability that at least 1520 of the people in the sample find shopping frustrating when $n = 2500$ and $p = 0.6$. We act as though the count X has the $N(1500, 24.4949)$ distribution. To obtain the normal approximation for this example, select **Calc ➤ Probability Distributions ➤ Normal** from the menu. In the dialog box, select Cumulative probability and specify a mean equal to 1500, a standard deviation equal to 24.4949, and 1520 for the input constant. As with the binomial distribution, the cumulative probability is $P(X \leq x)$. To calculate $P(X \geq x)$, we must subtract the result from 1. As we see from the following output, the normal approximation gives $P(X \leq 1520) = 0.7929$, so $P(X \geq 1520) \approx 0.21$, approximately the same as the exact results we obtained previously.

Cumulative Distribution Function

```
Normal with mean = 1500 and standard deviation = 24.4949

   x  P( X <= x )
1520     0.792892
```

EXERCISES

12.5 If the parents in Example 12.4 have 5 children, the number who have type O blood is a random variable X that has the binomial distribution with $n = 5$ and $p = 0.25$.

(a) What are the possible values of X?

(b) Select **Calc ➤ Probability Distributions ➤ Binomial** from the menu to find the probability of each value of X. Select **Graph ➤ Bar Chart** from the menu to display this distribution. (Because probabilities are long-run proportions, a graph with the probabilities as the heights of the bars shows what the distribution of X would be in very many repetitions.)

(c) Select **Calc ➤ Calculator** from the menu to find the mean and standard deviation of the number of children with type O blood. Mark the location of the mean on the graph you made in part (b).

12.6 When an opinion poll or telemarketer calls residential telephone numbers at random, 20% of the calls reach a live person. You watch the random dialing machine make 15 calls. The number that reach a person has the binomial distribution with $n = 15$ and $p = 0.2$. Select **Calc ➤ Probability Distributions ➤ Binomial** from the menu to answer the following questions.

(a) What is the probability that exactly 3 calls reach a person?

(b) What is the probability that 3 or fewer calls reach a person?

12.11 According to Benford's law (Example 9.5 in BPS) the probability that the first digit of the amount of a randomly chosen invoice is a 1 or a 2 is 0.477. You examine 90 invoices from a vendor and find that 29 have first digits 1 or 2. If Benford's law holds, the count of 1s and 2s will have the binomial distribution with $n = 90$ and $p = 0.477$. Too few 1s and 2s suggests fraud. Select **Calc ➤ Probability Distributions ➤ Binomial** from the menu to find the probability of 29 or fewer if the invoices follow Benford's law. Do you suspect that the invoice amounts are not genuine?

12.12 In 1998, Mark McGwire of the St. Louis Cardinals hit 70 home runs, a new major league record. Was this feat as surprising as most of us thought? In the three seasons before 1998, McGwire hit a home run in 11.6% of his times at bat. He went to bat 509 times in 1998. McGwire's home run count in 509 times at bat has approximately the binomial distribution with $n = 509$ and $p = 0.116$. Select **Calc ➤ Calculator** from the menu to find the mean and standard deviation for number of home runs he will hit in 509 times at bat. Select **Calc ➤ Probability Distributions ➤ Normal** to find the approximate probability of 70 or more home runs.

12.13 Checking for survey errors. One way of checking the effect of undercoverage, nonresponse, and other sources of error in a sample survey is to compare the sample with known facts about the population. About 12% of American adults are black. The number X of blacks in a random sample of 1500 adults should therefore vary with the binomial ($n = 1500$, $p = 0.12$) distribution.

(a) Select **Calc ➤ Calculator** from the menu to find the mean and standard deviation of X.

(b) Select **Calc ➤ Probability Distributions ➤ Normal** to find the approximate probability that the sample will contain 170 or fewer blacks. Be sure to check that you can safely use the approximation.

12.16 Each entry in a table of random digits like Table B has probability 0.1 of being a 0, and digits are independent of each other.

(a) Select **Calc ➤ Probability Distributions ➤ Binomial** to find the probability that a group of five digits from the table will contain at least one 0.

(b) Select **Calc ➤ Calculator** to find the mean number of 0s in lines 40 digits long.

12.17 People with type O-negative blood are universal donors whose blood can be safely given to anyone. Only 7.2% of the population have O-negative blood. A blood center is visited by 20 donors in an afternoon. Select **Calc ➤ Probability Distributions ➤ Binomial** to find the probability that there are at least two universal donors among them.

12.18 In a test for ESP (extrasensory perception), a subject is told that cards the experimenter can see but he cannot contain either a star, a circle, a wave, or a square. As the experimenter looks at each of 20 cards in turn, the subject names the shape on the card. A subject who is just guessing has probability 0.25 of guessing correctly on each card.

(a) The count of correct guesses in 20 cards has a binomial distribution. What are n and p?

(b) What is the mean number of correct guesses in many repetitions?

(c) Select **Calc ➤ Probability Distributions ➤ Binomial** to find the probability of exactly 5 correct guesses.

12.19 A believer in the "random walk" theory of stock markets thinks that an index of stock prices has probability 0.65 of increasing in any year. Moreover, the change in the index in any given year is not influenced by whether it rose or fell in earlier years. Let X be the number of years among the next 5 years in which the index rises.

(a) X has a binomial distribution. What are n and p?

(b) What are the possible values that X can take?

(c) Enter the values 0 through 5 into a column in your worksheet. Select **Calc ➤ Probability Distributions ➤ Binomial** to find the probability of each value of X. Select **Graph ➤ Chart** to draw a graph for the distribution of X. Make sure you enter your X values under X and your probabilities under Y.

(d) Select **Calc ➤ Calculator** from the menu to find the mean and standard deviation of this distribution. Mark the location of the mean on your histogram.

12.20 Twenty percent of American households own three or more motor vehicles. You choose 12 households at random.

(a) Select **Calc ➤ Probability Distributions ➤ Binomial** to find the probability that none of the chosen households owns three or more vehicles. What is the probability that at least one household owns three or more vehicles?

(b) Select **Calc ➤ Calculator** from the menu to find the mean and standard deviation of the number of households in your sample that own three or more vehicles.

(c) Select **Calc ➤ Probability Distributions ➤ Binomial** to find the probability that your sample count is greater than the mean.

12.22 Here is a simple probability model for multiple-choice tests. Suppose that each student has probability p of correctly answering a question chosen at random from a universe of possible questions. (A strong student has a higher p than a weak student.) Answers to different questions are independent. Jodi is a good student for whom $p = 0.75$.

(a) Select **Calc ➤ Calculator** from the menu to find the mean and standard deviation of Jodi's score on a 100-question test. Select **Calc ➤ Probability Distributions ➤ Normal** to find the approximate probability that Jodi scores 70% or lower on a 100-question test.

(b) If the test contains 250 questions, what is the probability that Jodi will score 70% or lower?

12.24 According to the Census Bureau, 9.96% of American adults (age 18 and over) are Hispanics. An opinion poll plans to contact an SRS of 1200 adults.

(a) Select **Calc ➤ Calculator** from the menu to find the mean and standard deviation of the mean number of Hispanics in such samples.

(b) Select **Calc ➤ Probability Distributions ➤ Normal** to find the approximate probability that the sample will contain fewer than 100 Hispanics.

12.25 Leakage from underground gasoline tanks at service stations can damage the environment. It is estimated that 25% of these tanks leak. You examine 15 tanks chosen at random, independently of each other.

(a) Select **Calc ➤ Calculator** from the menu to find the mean number of leaking tanks in such samples of 15.

(b) Select **Calc ➤ Probability Distributions ➤ Binomial** to find the probability that 10 or more of the 15 tanks leak.

(c) Now you do a larger study, examining a random sample of 1000 tanks nationally. What is the probability that at least 275 of these tanks are leaking?

12.27 High school dropouts make up 13% of all Americans aged 18 to 24. A vocational school that wants to attract dropouts mails an advertising flyer to 25,000 persons between the ages of 18 and 24.

(a) If the mailing list can be considered a random sample of the population, what is the mean and standard deviation for the number of high school dropouts who out will receive the flyer? Select **Calc ➤ Calculator** from the menu to find the answer.

(b) Select **Calc ➤ Probability Distributions ➤ Normal** to find the approximate probability that at least 3500 dropouts will receive the flyer.

12.28 While he was a prisoner of the Germans during World War II, John Kerrich tossed a coin 10,000 times. He got 5067 heads. Take Kerrich's tosses to be an SRS from the population of all possible tosses of his coin. If the coin is perfectly balanced, $p = 0.5$. Is there reason to think that Kerrich's coin gave too many heads to be balanced? To answer this question, select **Calc ➤ Calculator** from the menu to find the mean and standard deviation for number of heads in 10,000 coin tosses. Select **Calc ➤ Probability Distributions ➤ Normal** to find the approximate probability that a balanced coin would give 5067 or more heads in 10,000 tosses. What do you conclude?

12.29 Example 12.5 in BPS concerns the count of bad switches in inspection samples of size 10. The count has the binomial distribution with $n = 10$ and $p = 0.1$. The example calculates that the probability of getting a sample with exactly 1 bad switch is 0.3874. Of course, when we inspect only a few lots, the proportion of samples with exactly 1 bad switch will differ from this probability. Select **Calc ➤ Random Data ➤ Binomial** from the menu to simulate inspecting 20 lots. Record the number of bad switches (the count of heads) in each of the 20 samples. What proportion of the 20 lots had exactly 1 bad switch? Remember that probability tells us only what happens in the long run.

Chapter 13
Confidence Intervals: The Basics

Topic to be covered in this chapter:

One-Sample Z Confidence Interval

One-Sample Z Confidence Interval

Confidence intervals for the population mean μ, with σ known, can be calculated by selecting

Stat ➤ Basic Statistics ➤ 1-Sample Z

from the menu. This interval goes from $\bar{x} - z^*\left(\sigma/\sqrt{n}\right)$ to $\bar{x} + z^*\left(\sigma/\sqrt{n}\right)$ where $\bar{x}$ is the mean of the data, n is the sample size, and z^* is the critical value from the normal table corresponding to the confidence level.

Example 13.13 of BPS describes a manufacturer of pharmaceutical products. The laboratory verifies the concentration of active ingredients by analyzing each specimen three times. The standard deviation of this distribution is known to be $\sigma = 0.0068$ grams per liter. Three analyses of one specimen give concentrations

0.8403 0.8363 0.8447

We want a 99% confidence interval for the true concentration σ. We enter the data into column C1 of a Minitab worksheet and then calculate the confidence interval. In the following dialog box, we enter the column containing the variable that you want to calculate the confidence interval for and enter a value for σ in the Standard deviation box. As an alternative, we could enter the mean and sample size instead of the column containing the data.

1-Sample Z (Test and Confidence Interval)

Samples in columns: concentration

Summarized data
Sample size:
Mean:

Standard deviation: .0068

Test mean: (required for test)

Select   Graphs...   Options...

Help   OK   Cancel

Click on the Options button. In the Options subdialog box that appears, specify a Confidence level and click OK.

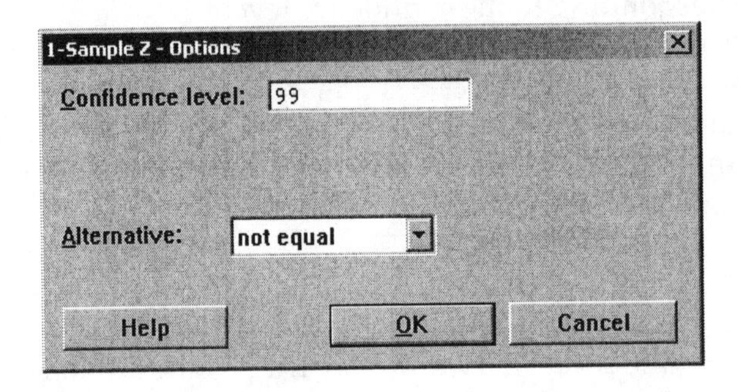

One-Sample Z: concentration

```
The assumed standard deviation = 0.0068


Variable        N      Mean     StDev   SE Mean          99% CI
concentration   3  0.840433  0.004202  0.003926  (0.830321, 0.850546)
```

Note that we specified a 99% interval by entering 99 in the Options subdialog box. The values for Mean and StDev listed with the confidence intervals are the same as those that would be obtained by selecting **Stat ➤ Basic Statistics ➤ Display Descriptive Statistics** from the menu. The value given for SE Mean is calculated with the known value of σ as follows.

$$\frac{\sigma}{\sqrt{n}} = \frac{.0068}{\sqrt{3}} = .00393.$$

Minitab can be used to find the critical value that is used for a specific level of confidence. We can select **Calc ➤ Probability Distributions ➤ Normal** from the menu to find the value of *z* that has a specific area below it. For a level *C* confidence interval, we want to have an area of (1-C)/2 above and $1-\{(1-C)/2\}$ below the critical value. If we want the critical value for a 75% level of confidence, a value not included in Table C, we let *C* = 0.75. Therefore, $1-\{(1-C)/2\}$ = 0.875. Select **Calc ➤ Probability Distributions ➤ Normal** from the menu. In the dialog box, select Inverse cumulative probability and enter Mean = 0 and Standard deviation = 1.0. Enter 0.875 as the Input constant and click OK.

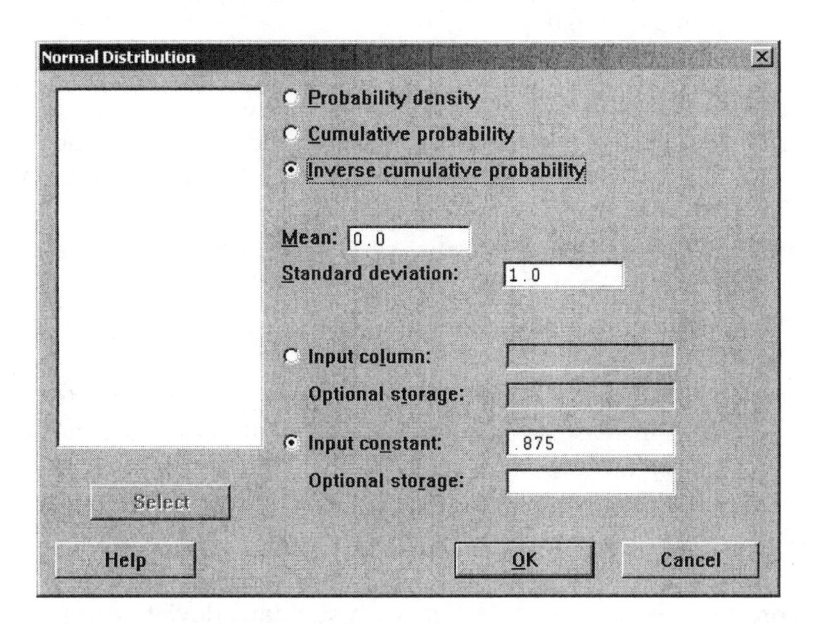

Inverse Cumulative Distribution Function

```
Normal with mean = 0 and standard deviation = 1

P( X <= x )          x
      0.875    1.15035
```

Therefore, for a 75% confidence interval, $z^* = 1.15035$.

EXERCISES

13.4 Select **Calc ➤ Probability Distributions ➤ Normal** from the menu to find the critical value z^* for confidence level 97.5%. Start by making a copy of Figure 13.5 with C = 0.975 that shows how much area is left in each tail when the central area is 0.975.

13.6 Here and in EX13-06.MTW are the IQ test scores of 31 seventh-grade girls in a Midwest school district:

114	100	104	89	102	91	114	114	103	105	108
130	120	132	111	128	118	119	86	72	111	103
74	112	107	103	98	96	112	112	93		

(a) We expect the distribution of IQ scores to be close to Normal. Select **Graph ➤ Stem-and-Leaf** from the menu to make a stemplot of the distribution of these 31 scores (split the stems) to verify that there are no major departures from Normality.

(b) Treat the 31 girls as an SRS of all seventh-grade girls in the school district. Suppose that the standard deviation of IQ scores in this population is known to be $\sigma = 15$. Select **Stat ➤ Basic Statistics ➤ 1-Sample Z** to give a 99% confidence interval for the mean score in the population.

13.7 Examples 13.3 and 13.4 give confidence intervals for the concentration μ based on 3 measurements with $\bar{x} = 0.8404$ and $\sigma = 0.0068$. The 99% confidence interval is 0.8303 to 0.8505 and the 90% confidence interval is 0.8339 to 0.8469.

(a) Select **Stat ➤ Basic Statistics ➤ 1-Sample Z** to find the 80% confidence interval for μ.

(b) Select **Stat ➤ Basic Statistics ➤ 1-Sample Z** to find the 99.9% confidence interval for μ.

(c) Make a sketch like Figure 13.7 to compare all four intervals. How does increasing the confidence level affect the length of the confidence interval?

13.8 High school students who take the SAT mathematics exam a second time generally score higher than they did on their first try. The change in score has a Normal distribution with standard deviation $\sigma = 50$. A random sample of 1000 students gains an average of $\bar{x} = 22$ points on their second try.

(a) Select **Stat ➤ Basic Statistics ➤ 1-Sample Z** to give a 95% confidence interval for the mean score gain μ in the population of all students.

(b) Select **Stat ➤ Basic Statistics ➤ 1-Sample Z** to give the 90% and 99% confidence intervals for μ.

(c) What are the margins of error for 90%, 95%, and 99% confidence? How does increasing the confidence level affect the margin of error of a confidence interval?

13.9 Sample size and margin of error. A sample of 1000 high school students gained an average of $\bar{x} = 22$ points in their second attempt at the SAT mathematics exam. The change in score has a Normal distribution with standard deviation $\sigma = 50$.

(a) Select **Stat ➤ Basic Statistics ➤ 1-Sample Z** to give a 95% confidence interval for the mean score gain μ in the population of all students.

(b) Suppose that the same result, $\bar{x} = 22$, had come from a sample of 250 students. Select **Stat ➤ Basic Statistics ➤ 1-Sample Z** to give the 95% confidence interval for the population mean μ in this case.

(c) Then suppose that a sample of 4000 students had produced the sample mean $\bar{x} = 22$. Again select **Stat ➤ Basic Statistics ➤ 1-Sample Z** to give the 95% confidence interval for μ.

(d) What are the margins of error for samples of size 250, 1000, and 4000? How does increasing the sample size affect the margin of error of a confidence interval?

13.13 Breast-feeding mothers secrete calcium into their milk. Some of the calcium may come from their bones, so mothers may lose bone mineral. Researchers measured the percent change during three months of breast feeding of the mineral content in the spines of 47 mothers. Here and in EX13-13.MTW are the data:

−4.7	−2.5	−4.9	−2.7	−0.8	−5.3	−8.3	−2.1	−6.8	−4.3
2.2	−7.8	−3.1	−1.0	−6.5	−1.8	−5.2	−5.7	−7.0	−2.2
−6.5	−1.0	−3.0	−3.6	−5.2	−2.0	−2.1	−5.6	−4.4	−3.3
−4.0	−4.9	−4.7	−3.8	−5.9	−2.5	−0.3	−6.2	−6.8	1.7
0.3	−2.3	0.4	−5.3	0.2	−2.2	−5.1			

(a) Select **Graph ➤ Stem-and-Leaf** from the menu to make a stemplot of the data. The data appear to follow a Normal distribution quite closely.

(b) Suppose that the percent change in the population of all nursing mothers has standard deviation $\sigma = 2.5\%$. Select **Stat ➤ Basic Statistics ➤ 1-Sample Z** to give a 99% confidence interval for the mean percent change in the population.

13.14 Biologists studying the healing of skin wounds measured the rate at which new cells closed a razor cut made in the skin of an anesthetized newt. Here and in EX13-14.MTW are data from 18 newts, measured in micrometers (millionths of a meter) per hour:

29	27	34	40	22	28	14	35	26
35	12	30	23	18	11	22	23	33

(a) Select **Graph ➤ Stem-and-Leaf** from the menu to make a stemplot of the healing rates (split the stems). It is diffcult to assess Normality from 18 observations, but look for outliers or extreme skewness. What do you find?

(b) Scientists usually assume that animal subjects are SRSs from their species or genetic type. Treat these newts as an SRS and suppose you know that the standard deviation of healing rates for this species of newt is 8 micrometers per hour. Select **Stat ➤ Basic Statistics ➤ 1-Sample Z** to give a 90% confidence interval for the mean healing rate for the species.

(c) A friend who knows almost no statistics follows the formula $\bar{x} \pm 1.96\,\sigma/\sqrt{n}$ in a biology lab manual to get a 95% confidence interval for the mean. Is her interval wider or narrower than yours?

Explain to her why it makes sense that higher confidence changes the length of the interval.

13.15 Here and in EX13-15.MTW are data from students doing a laboratory exercise: how heavy a load (pounds) is needed to pull apart pieces of Douglas fir 4 inches long and 1.5 inches square.

33,190	31,860	32,590	26,520	33,280
32,320	33,020	32,030	30,460	32,700
23,040	30,930	32,720	33,650	32,340
24,050	30,170	31,300	28,730	31,920

(a) Suppose that the strength of pieces of wood like these follows a Normal distribution with standard deviation 3000 pounds. Select **Stat ➤ Basic Statistics ➤ 1-Sample Z** to give a 90% confidence interval for the mean load required to pull the wood apart.

(b) We are willing to regard the wood pieces prepared for the lab session as an SRS of all similar pieces of Douglas fir. Engineers also commonly assume that characteristics of materials vary Normally. Make a graph to show the shape of the distribution for these data. Does the Normality assumption appear safe?

13.16 Sulfur compounds cause "off-odors" in wine, so winemakers want to know the odor threshold, the lowest concentration of a compound that the human nose can detect. The odor threshold for dimethyl sulfide (DMS) in trained wine tasters is about 25 micrograms per liter of wine (μg/L). The untrained noses of consumers may be less sensitive, however. Here are the DMS odor thresholds for 10 untrained students:

31 31 43 36 23 34 32 30 20 24

Assume that the standard deviation of the odor threshold for untrained noses is known to be $\sigma = 7 \mu g/L$.

(a) Select **Graph ➤ Stem-and-Leaf** from the menu to make a stemplot to verify that the distribution is roughly symmetric with no outliers. (More data confirm that there are no systematic departures from Normality.)

(b) Select **Stat ➤ Basic Statistics ➤ 1-Sample Z** to give a 95% confidence interval for the mean DMS odor threshold among all students.

13.17 Here and in EX13-17.MTW are measurements (in millimeters) of a critical dimension on a sample of auto engine crankshafts:

224.120	224.001	224.017	223.982	223.989	223.961
223.960	224.089	223.987	223.976	223.902	223.980
224.098	224.057	223.913	223.999		

The data come from a production process that is known to have standard deviation $\sigma = 0.060$ mm. The process mean is supposed to be $\mu = 224$ mm but can drift away from this target during production.

(a) We expect the distribution of the dimension to be close to Normal. Make a stemplot or histogram of these data and describe the shape of the distribution.

(b) Select **Stat ➤ Basic Statistics ➤ 1-Sample Z** to give a 95% confidence interval for the process mean at the time these crankshafts were produced.

13.18 A class survey in a large class for first-year college students asked, "About how many minutes do you study on a typical weeknight?" The mean response of the 269 students was $\bar{x} = 137$ minutes. Suppose we know that the study time follows a Normal distribution with standard deviation $\sigma = 65$ minutes in the population of all first-year students at this university.

(a) Use the survey result to give a 99% confidence interval for the mean study time of all first-year students. Select **Stat ➤ Basic Statistics ➤ 1-Sample Z** to calculate the interval.

(b) What condition not yet mentioned is needed for your confidence interval to be valid?

13.19 Table I.1 in BPS and EX13-19.MTW give data on 38 consecutive patients who came to a medical center for treatment of Hallux abducto valgus (HAV), a deformation of the big toe. It is reasonable to consider these patients as an SRS of people suffering from HAV. The seriousness of the deformity is measured by the angle (in degrees) of deformity.

(a) The data contain one high outlier. What is the angle for this outlier? The presence of the outlier violates the conditions for our confidence interval. Suppose that there is a good medical reason to remove the outlier.

(b) The remaining 37 observations follow a Normal distribution closely. Assume that angle has a Normal distribution with standard deviation $\sigma = 6.3$ degrees. Select **Stat ➤ Basic Statistics ➤ 1-Sample Z** to give a 95% confidence interval for the mean angle of deformity in the population.

Chapter 14
Tests of Significance: The Basics

Topic to be covered in this chapter:

One-Sample Z Test

One-Sample *Z* Test

As with confidence intervals, we can do a hypothesis test for a population mean μ, with σ known, by selecting

Stat ➤ Basic Statistics ➤ 1-Sample Z

from the menu. In the dialog box, choose Samples in columns if you have raw data or choose Summarized data if you have summary values for the sample size and mean. Enter a value for the population standard deviation in the Standard deviation box. In the Test mean box, specify the null hypothesis test value and click on OK.

In Example 14.10 of BPS, we want to determine if there is significant evidence at the 1% level that the true concentration is not 0.86%. This calls for a test of the hypothesis that $\mu = 0.86$ against the alternative $\mu \neq 0.86$.

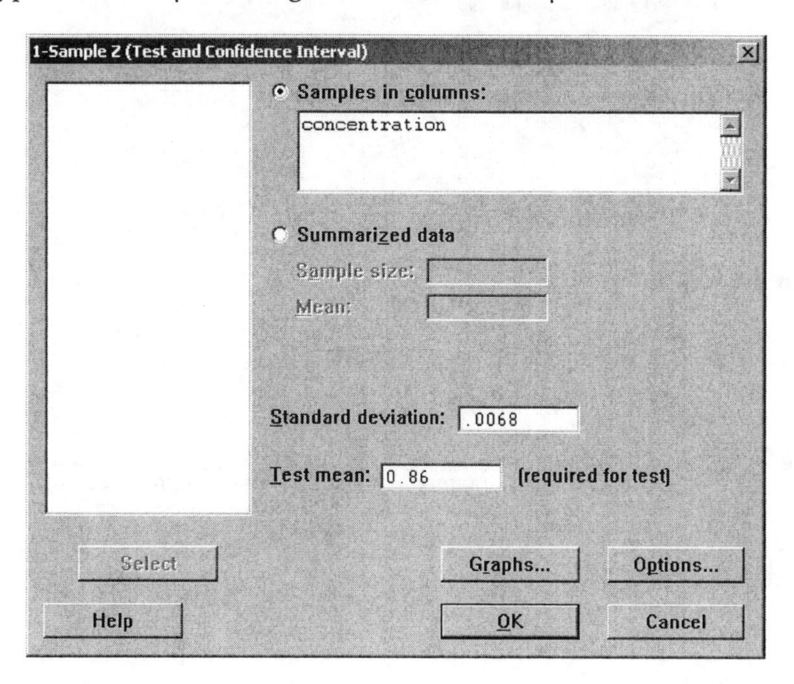

Click on the Options button. In the Options subdialog box, specify the alternative hypothesis. You can choose "less than" (lower-tailed), "not equal" (two-tailed), or "greater than" (upper-tailed) and click OK.

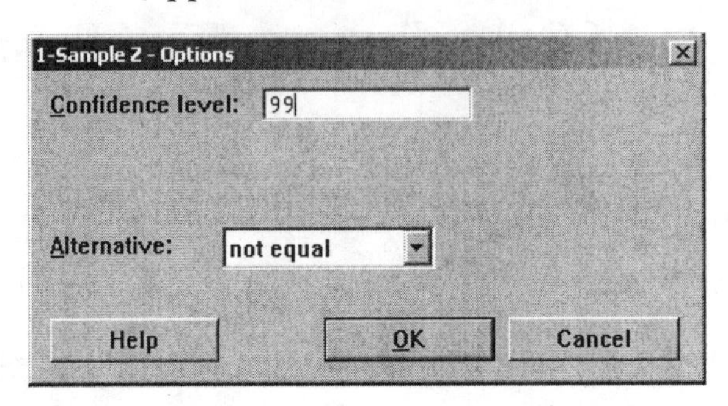

One-Sample Z: concentration

```
Test of mu = 0.86 vs not = 0.86
The assumed standard deviation = 0.0068


Variable       N      Mean     StDev   SE Mean            99% CI               Z
concentration  3  0.840433  0.004202  0.003926  (0.830321, 0.850546)  -4.98

Variable           P
concentration  0.000
```

If instead we wish to test the hypothesis that $\mu = 0.86$ against the alternative $\mu < 0.86$ for the data given in Example 14.10 of BPS, we can choose 'less than' in the Options subdialog box.

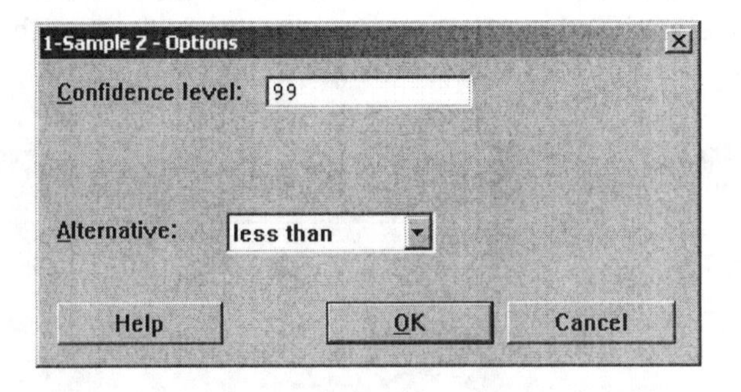

One-Sample Z: concentration

```
Test of mu = 0.86 vs < 0.86
The assumed standard deviation = 0.0068

                                                      99%
                                                    Upper
Variable       N      Mean     StDev   SE Mean      Bound      Z      P
concentration  3  0.840433  0.004202  0.003926  0.849567  -4.98  0.000
```

The *P*-value given is always smaller for the one-sided test. In fact, it is equal to half the *P*-value computed for the two-sided test. In both tests of Example 14.10, the *P*-value is 0 and the null hypothesis should be rejected.

Minitab can be used to find the critical value that would be required to reject the null hypothesis at a particular significance level. Select **Calc ➤ Probability Distributions ➤ Normal** from the menu to calculate a value associated with an area. We can find the value of z that has a specific area below it. Suppose that we wish to use a 1% significance criterion for Example 14.10 of BPS. If the test is two-sided, the probability in each tail must be 0.005 or less, so we find the value of z that has area 0.005 below it. In the dialog box, select Inverse cumulative probability and enter Mean = 0 and Standard deviation = 1.0. Enter 0.01 as the Input constant and click OK.

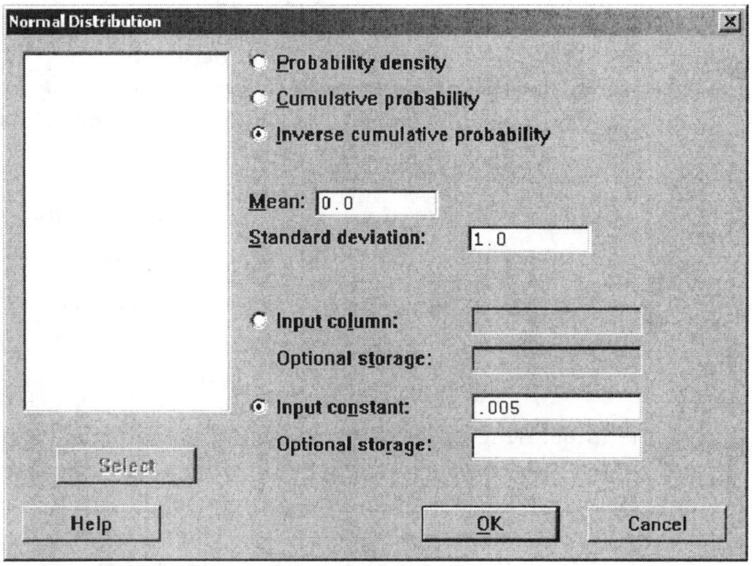

Normal Distribution

Probability density
Cumulative probability
Inverse cumulative probability

Mean: 0.0
Standard deviation: 1.0

Input column:
Optional storage:
Input constant: .005
Optional storage:

Select

Help OK Cancel

Inverse Cumulative Distribution Function

```
Normal with mean = 0 and standard deviation = 1.00000

P( X <= x )          x
    0.0050      -2.5758
```

Therefore, we would reject the null hypothesis at the 1% level of significance if $z \leq -2.5758$ or if $z \geq 2.5758$.

EXERCISES

14.17 An environmentalist group collects a liter of water from each of 45 random locations along a stream and measures the amount of dissolved oxygen in each specimen. The mean is 4.62 milligrams (mg). Select **Stat ➤ Basic Statistics ➤ 1-Sample Z** from the menu to see if this is strong evidence that the stream has a mean oxygen content of less than 5 mg per liter. (Suppose we know that dissolved oxygen varies among locations according to a Normal distribution with $\sigma = 0.92$ mg.)

14.18 We suspect that on the average students will score higher on their second attempt at the SAT mathematics exam than on their first attempt. Suppose we know that the changes in score (second try minus first try) follow a Normal distribution with standard deviation $\sigma = 50$. Here and in EX14-18.MTW are the results for 46 randomly chosen high school students:

−30	24	47	70	−62	55	−41	−32	128	−11
−43	122	−10	56	32	−30	−28	−19	1	17
57	−14	−58	77	27	−33	51	17	−67	29
94	−11	2	12	−53	−49	49	8	−24	96
120	2	−33	−2	−39	99				

Do these data give good evidence that the mean change in the population is greater than zero? State the hypotheses. Select **Stat ➤ Basic Statistics ➤ 1-Sample Z** from the menu to calculate a test statistic and its *P*-value. State your conclusion.

14.19 Here and in EX14-19.MTW are measurements (in millimeters) of a critical dimension on a sample of automobile engine crankshafts:

224.120	224.001	224.017	223.982	223.989	223.961
223.960	224.089	223.987	223.976	223.902	223.980
224.098	224.057	223.913	223.999		

The manufacturing process is known to vary Normally with standard deviation $\sigma = 0.060$ mm. The process mean is supposed to be 224 mm. Do these data give evidence that the process mean is not equal to the target value 224 mm? State the hypotheses. Select **Stat ➤ Basic Statistics ➤ 1-Sample Z** from the menu to calculate a test statistic and its *P*-value. Are you convinced that the process mean is not 224 mm?

14.22 A random number generator is supposed to produce random numbers that are uniformly distributed on the interval from 0 to 1. If this is true, the numbers generated come from a population with $\mu = 0.5$ and $\sigma = 0.2887$. A command to generate 100 random numbers gives outcomes with mean $\bar{x} = 0.4365$. Assume that the population σ remains fixed. We want to test

$$H_0: \mu = 0.5$$
$$H_a: \mu \neq 0.5$$

(a) Select **Stat ➤ Basic Statistics ➤ 1-Sample Z** from the menu to calculate a test statistic and its *P*-value.

(b) Is the result significant at the 5% level ($\alpha = 0.05$)?

(c) Is the result significant at the 1% level ($\alpha = 0.01$)?

14.23 Exercise 13.13 and EX14-23.MTW give the percent change in the mineral content of the spine for 47 mothers during three months of nursing a baby. As in that exercise, suppose that the percent change in the population of all nursing mothers has a Normal distribution with standard deviation $\sigma = 2.5\%$. Do these data give good evidence that on the average nursing mothers lose bone mineral? State the hypotheses. Select **Stat ➤ Basic Statistics ➤ 1-Sample Z** from the menu to calculate the z test statistic and P-value. What do you conclude?

14.26 A student group claims that first-year students at a university must study 2.5 hours per night during the school week. A skeptic suspects that they study less than that on the average. A class survey finds that the average study time claimed by 269 students is $\bar{x} = 137$ minutes. Regard these students as a random sample of all first-year students and suppose we know that study times follow a Normal distribution with standard deviation 65 minutes. Select **Stat ➤ Basic Statistics ➤ 1-Sample Z** from the menu to carry out a test of H_0: $\mu = 150$ against H_a: $\mu < 150$. What do you conclude?

14.27 Exercise 13.6 and EX14-27.MTW give the IQ test scores of 31 seventh-grade girls in a Midwest school district. IQ scores follow a Normal distribution with standard deviation $\sigma = 15$. Treat these 31 girls as an SRS of all seventh-grade girls in this district. IQ scores in a broad population are supposed to have mean $\mu = 100$. Is there evidence that the mean in this district differs from 100? State the hypotheses. Select **Stat ➤ Basic Statistics ➤ 1-Sample Z** from the menu to find the test statistic and its P-value. State your conclusion.

14.28 Sulfur compounds cause "off-odors" in wine, so winemakers want to know the odor threshold, the lowest concentration of a compound that the human nose can detect. The odor threshold for dimethyl sulfide (DMS) in trained wine tasters is about 25 micrograms per liter of wine (μg/L). The untrained noses of consumers may be less sensitive, however. Here and in EX14-28.MTW are the DMS odor thresholds for 10 untrained students:

31 31 43 36 23 34 32 30 20 24

Assume that the odor threshold for untrained noses is Normal with $\sigma = 7\mu$g/l. Select **Stat ➤ Basic Statistics ➤ 1-Sample Z** from the menu to see if there is evidence that the mean threshold for untrained tasters is less than 25 μg/L.

14.29 Exercise 13.14 and EX14-20.MTW give data and information about the rate at which skin wounds heal in newts. A newt expert says that 25 micrometers per hour is the usual rate and that the standard deviation of healing rates for this species of newt is 8 micrometers per hour. Select **Stat ➤ Basic Statistics ➤ 1-Sample Z** from the menu to see if the data give evidence against this claim.

14.48 In Exercise 13.15 and EX14-48.MTW are data from students doing a laboratory exercise to see how heavy a load (pounds) is needed to pull apart pieces of Douglas fir 4 inches long and 1.5 inches square. Select **Stat ➤ Basic Statistics ➤ 1-Sample Z** from the menu to find the 90% confidence interval for the mean load required to pull apart pieces of Douglas fir. Use this interval to answer these questions:

(a) Is there significant evidence at the $\alpha = 0.10$ level against the hypothesis that the mean is 32,000 pounds against the two-sided alternative?

(b) Is there significant evidence at the $\alpha = 0.10$ level against the hypothesis that the mean is 31,500 pounds against the two-sided alternative?

Chapter 15
Inference in Practice

Topic to be covered in this chapter:

Calculating Power

Calculating Power

Minitab can be used to calculate the power for a one-sample Z-test. Select

Stat ➤ Power and Sample Size ➤ 1-Sample Z

from the menu. In the dialog box, enter the acceptable values for any two of these parameters and Minitab will solve for the third. In the Standard deviation box, enter σ for your data.

Example 15.5 of BPS describes the power calculation for a one-sided Z-test for a cola maker. The cola maker determines that a sweetness loss is too large to accept if the mean response for all tasters is $\mu = 1.1$. Consider a test of

$$H_0: \mu = 0$$
$$H_a: \mu > 0$$

at the 5% level of significance. We want to find the power of the test against the specific alternative $\mu = 1.1$. In the dialog box, enter the sample size $n = 10$, the difference = 1.1, and $\sigma = 1$.

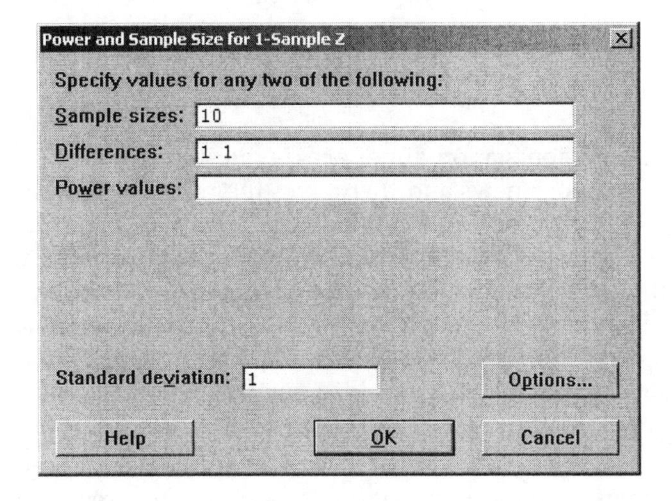

Click on the Options button. In the Options subdialog box, check the Greater than hypothesis, enter Significance level = 0.05, and click OK.

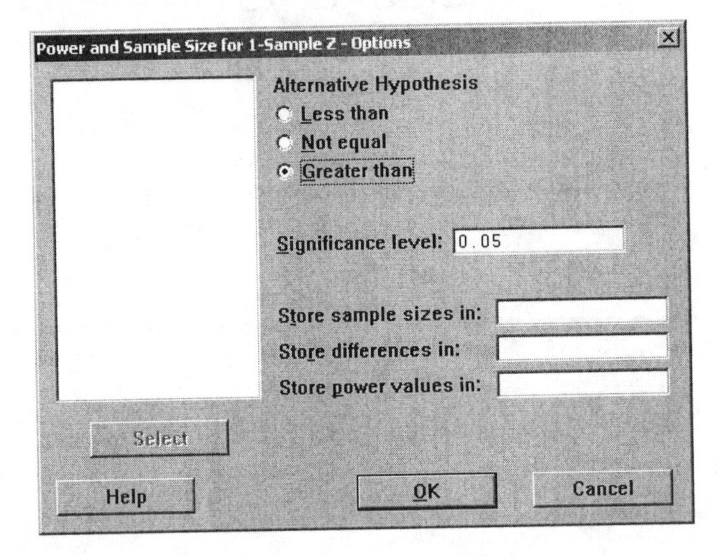

As shown below, the power is close to 0.97. Therefore, we can be quite confident that the test will reject H_0 when this alternative is true.

Power and Sample Size

```
1-Sample Z Test

Testing mean = null (versus > null)
Calculating power for mean = null + difference
Alpha = 0.05  Assumed standard deviation = 1


              Sample
Difference      Size     Power
       1.1        10  0.966647
```

EXERCISES

15.1 A local television station announces a question for a call-in opinion poll on the six o'clock news and then gives the response on the eleven o'clock news. Today's question is "What yearly pay do you think members of the City Council should get? Call us with your number." In all, 958 people call. The mean pay they suggest is $\bar{x}$ = \$8740 per year, and the standard deviation of the responses is s = \$1125. For a large sample such as this, s is very close to the unknown population σ, so take σ = \$1125. The station calculates the 95% confidence interval for the mean pay μ that all citizens would propose for council members to be \$8669 to \$8811.

(a) Select **Stat ➤ Basics Statistics ➤ 1-Sample Z** and enter the summarized data. Is the station's calculation correct?

(b) Does their conclusion describe the population of all the city's citizens? Explain your answer.

15.2 A survey of licensed drivers inquired about running red lights. One question asked, "Of every ten motorists who run a red light, about how many do you think will be caught?" The mean result for 880 respondents was $\bar{x} = 1.92$ and the standard deviation was $s = 1.83$. For this large sample, s will be close to the population standard deviation σ, so suppose we know that $\sigma = 1.83$.

(a) Select **Stat ➤ Basics Statistics ➤ 1-Sample Z** to give a 95% confidence interval for the mean opinion in the population of all licensed drivers.

(b) The distribution of responses is skewed to the right rather than Normal. This will not strongly affect the z confidence interval for this sample. Why not?

(c) The 880 respondents are an SRS from completed calls among 45,956 calls to randomly chosen residential telephone numbers listed in telephone directories. Only 5029 of the calls were completed. This information gives two reasons to suspect that the sample may not represent all licensed drivers. What are these reasons?

15.4 An interviewer asks, "How much do you plan to spend for gifts this holiday season?" of 250 customers at a large shopping mall. The sample mean and standard deviation of the responses are $\bar{x} = \$237$ and $s = \$65$.

(a) The distribution of spending is skewed, but we can act as though $\bar{x}$ is Normal. Why?

(b) For this large sample, we can act as if $\sigma = \$65$ because the sample s will be close to the population σ. Select **Stat ➤ Basics Statistics ➤ 1-Sample Z** to give a 99% confidence interval for the mean gift spending of all adults.

(c) This confidence interval can't be trusted because the sample responses may be badly biased. Suggest some reasons why the responses may be biased.

15.5 Suppose that in the absence of special preparation SAT mathematics (SATM) scores vary Normally with mean $\mu = 475$ and $\sigma = 100$. One hundred students go through a rigorous training program designed to raise their SATM scores by improving their mathematics skills. Carry out a test of

$$H_0: \mu = 475$$
$$H_a: \mu > 475$$

in each of the following situations:

(a) The students' average score is $\bar{x} = 491.4$. Select **Stat ➤ Basics Statistics ➤ 1-Sample Z** to see if this result is significant at the 5% level.

(b) The average score is $\bar{x} = 491.5$. Select **Stat ➤ Basics Statistics ➤ 1-Sample Z** to see if this result is significant at the 5% level. The difference between the two outcomes in (a) and (b) is of no importance. Beware attempts to treat $\alpha = 0.05$ as sacred.

15.6 Suppose that SAT mathematics scores in the absence of coaching vary Normally with mean $\mu = 475$ and $\sigma = 100$. Suppose also that coaching may change μ but does not change σ. An increase in an SAT score from 475 to 478 is of no importance in seeking admission to college, but this unimportant change can be statistically very significant. Select **Stat ➤ Basics Statistics ➤ 1-Sample Z** to calculate the *P*-value for the test of

$$H_0: \mu = 475$$
$$H_a: \mu > 475$$

in each of the following situations:

(a) A coaching service coaches 100 students. Their SATM scores average $\bar{x} = 478$.

(b) By the next year, the service has coached 1000 students. Their SATM scores average $\bar{x} = 478$.

(c) An advertising campaign brings the number of students coached to 10,000. Their average score is still $\bar{x} = 478$.

15.7 Select **Stat ➤ Basics Statistics ➤ 1-Sample Z** to give a 99% confidence interval for the mean SATM score μ after coaching in each part of the previous exercise. For large samples, the confidence interval tells us, "Yes, the mean score is higher than 475 after coaching but only by a small amount."

15.11 Bottles of a popular cola are supposed to contain 300 milliliters (ml) of cola. There is some variation from bottle to bottle because the filling machinery is not precise. The distribution of the contents is Normal with standard deviation $\sigma = 3$ ml. Will inspecting six bottles discover underfilling? The hypotheses are

$$H_0: \mu = 300$$
$$H_a: \mu < 300$$

A 5% significance test rejects H_0 if $z \leq 1.645$, where the test statistic z is

$$z = \frac{\bar{x} - 300}{3/\sqrt{6}}$$

Power calculations help us see how large a shortfall in the bottle contents the test can be expected to detect.

(a) Select **Stat ➤ Power and Sample Size ➤ 1-Sample Z** to find the power of this test against the alternative $\mu = 299$.

(b) Select **Stat ➤ Power and Sample Size ➤ 1-Sample Z** to find the power against the alternative $\mu = 295$.

(c) Is the power against $\mu = 290$ higher or lower than the value you found in (b)? (Don't actually calculate that power.) Explain your answer.

15.12 Increasing the sample size increases the power of a test when the level α is unchanged. Suppose that in the previous exercise a sample of n bottles had been measured. In that exercise, $n = 6$. The 5% significance test still rejects H_0 when $z \le -1.645$, but the z statistic is now

$$z = \frac{\bar{x} - 300}{3/\sqrt{n}}$$

(a) Select **Stat ➤ Power and Sample Size ➤ 1-Sample Z** to find the power of this test against the alternative $\mu = 299$ when $n = 25$.

(b) Select **Stat ➤ Power and Sample Size ➤ 1-Sample Z** to find the power of this test against the alternative $\mu = 299$ when $n = 100$.

15.14 You have the NAEP quantitative scores for an SRS of 840 young men. You plan to test hypotheses about the population mean score

$$H_0: \mu = 275$$
$$H_a: \mu < 275$$

at the 1% level of significance. The population standard deviation is known to be $\sigma = 60$. The z test statistic is

$$z = \frac{\bar{x} - 275}{60/\sqrt{840}}$$

(a) What is the rule for rejecting H_0 in terms of z?

(b) What is the probability of a Type I error?

(c) You want to know whether this test will usually reject H_0 when the true population mean is 270, 5 points lower than the null hypothesis claims. Answer this question by selecting **Stat ➤ Power and Sample Size ➤ 1-Sample Z** to find the power of this test against the alternative when $\mu = 270$.

15.22 A marketing consultant observes 50 consecutive shoppers at a supermarket. Here and in EX15-22.MTW are the amounts (in dollars) spent in the store by these shoppers.

3.11	8.88	9.26	10.81	12.69	13.78	15.23	15.62	17.00	17.39
18.36	18.43	19.27	19.50	19.54	20.16	20.59	22.22	23.04	24.47
24.58	25.13	26.24	26.26	27.65	28.06	28.08	28.38	32.03	34.98
36.37	38.64	39.16	41.02	42.97	44.08	44.67	45.40	46.69	48.65
50.39	52.75	54.80	59.07	61.22	70.32	82.70	85.76	86.37	93.34

(a) Why is it risky to regard these 50 shoppers as an SRS from the population of all shoppers at this store? Name some factors that might make 50 consecutive shoppers at a particular time unrepresentative of all shoppers.

(b) Select **Graph ➤ Stem-and-Leaf** from the menu to make a stemplot of the data. The stemplot suggests caution in using the z procedures for these data. Why?

15.31 Power calculations for two-sided tests follow the same outline as for one-sided tests. Example 14.10 of BPS presents a test of

$$H_0: \mu = 0.86$$
$$H_a: \mu \neq 0.86$$

at the 1% level of significance. The sample size is $n = 3$ and $\sigma = 0.0068$. We will find the power of this test against the alternative $\mu = 0.845$.

(a) Select **Stat ➤ Power and Sample Size ➤ 1-Sample Z** to find the power of this test against the alternative if the true mean is $\mu = 0.845$. This probability is the power.

(b) What is the probability that this test makes a Type II error when $\mu = 0.845$?

15.32 In Example 14.7 in BPS, a company medical director failed to find significant evidence that the mean blood pressure of a population of executives differed from the national mean $\mu = 128$. The medical director now wonders if the test used would detect an important difference if one were present. For the SRS of size 72 from a population with standard deviation $\sigma = 15$, the z statistic is

$$z = \frac{\bar{x} - 128}{15/\sqrt{72}}$$

The two-sided test rejects H_0: $\mu = 128$ at the 5% level of significance when $|z| \geq 1.96$.

(a) Select **Stat ➤ Power and Sample Size ➤ 1-Sample Z** to find the power of this test against the alternative $\mu = 134$.

(b) Select **Stat ➤ Power and Sample Size ➤ 1-Sample Z** to find the power of this test against $\mu = 122$. Can the test be relied on to detect a mean that differs from 128 by 6?

(c) If the alternative were farther from H_0, say $\mu = 136$, would the power be higher or lower than the values calculated in (a) and (b)?

Chapter 16
Inference About a Population Mean

Topics to be covered in this chapter:

One-Sample t Procedures
Matched pairs

One-Sample t Procedures

To compute a confidence interval and perform a hypothesis test of the mean when the population standard deviation σ is unknown, select

Stat ➤ Basic Statistics ➤ 1-Sample t

from the menu.

Example 16.1 of BPS describes a study in which biologists studying the healing of skin wounds measured the rate at which new cells closed a razor cut made in the skin of an anesthetized newt. Here and in EG16-01.MTW are data from 18 newts, measured in micrometers (millionths of a meter) per hour:

29	27	34	40	22	28	14	35	26
35	12	30	23	18	11	22	23	33

To find a 95% confidence interval for the mean choose **Stat ➤ Basic Statistics ➤ 1-Sample t** from the menu. In the dialog box shown on the following page, choose "Samples in columns" and enter the column containing the sample data. If you have summary values for the sample size, mean, and standard deviation, you can choose "Summarized data" and enter those values instead.

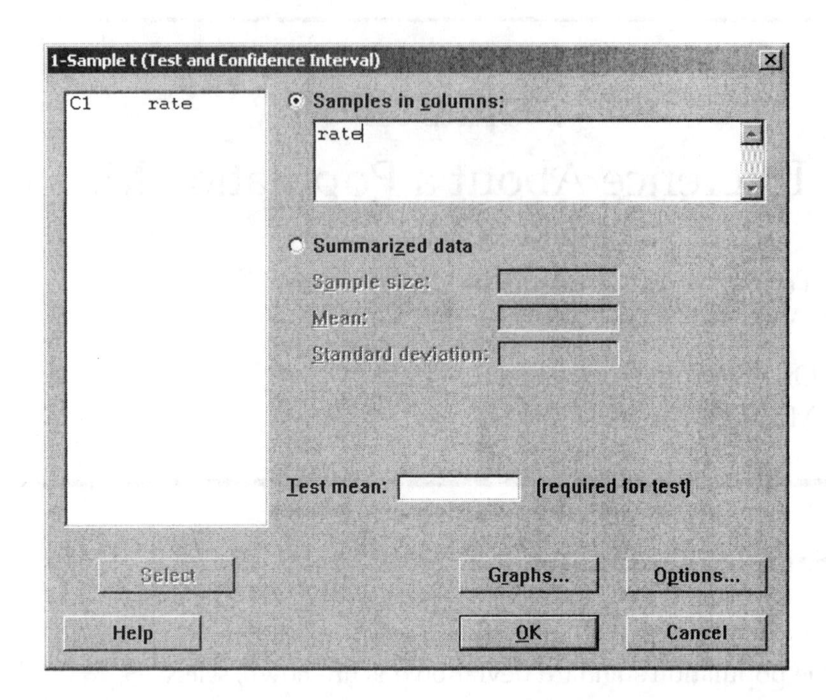

Click on the Options button to obtain the Options subdialog box. Select the level of confidence you want. For a confidence interval, make sure that the Alternative selected is "not equal".

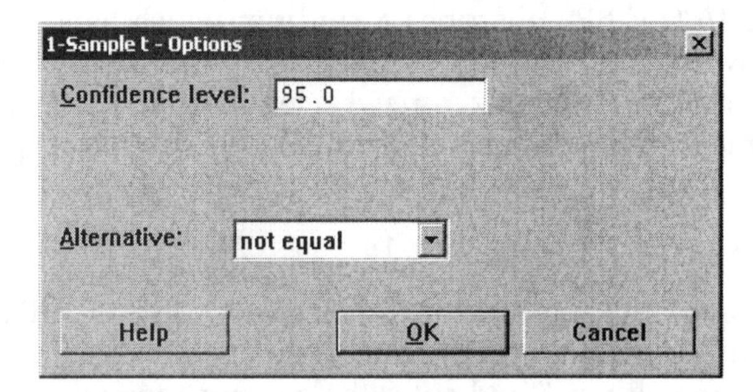

Minitab will calculated the confidence interval as

$$\bar{x} - t_{\alpha/2}\left(\frac{s}{\sqrt{n}}\right) \text{ to } \bar{x} + t_{\alpha/2}\left(\frac{s}{\sqrt{n}}\right)$$

where $\bar{x}$ is the mean of the data, s is the sample standard deviation, n is the sample size, and $t_{\alpha/2}$ is the critical value from a t-distribution with n–1 degrees of freedom. The 95% confidence interval for the healing rate is as follows.

One-Sample T: rate

```
Variable   N     Mean   StDev  SE Mean          95% CI
rate      18  25.6667  8.3243   1.9621  (21.5271, 29.8062)
```

If you click on the Graphs button in the 1-Sample t dialog box, you may choose to display a histogram, a plot of the individual values, and/or a boxplot of the data. A histogram is useful for checking whether the Normality assumption for inference has been satisfied. The histogram for rate is shown here. The graph shows the sample mean and the confidence interval for the mean.

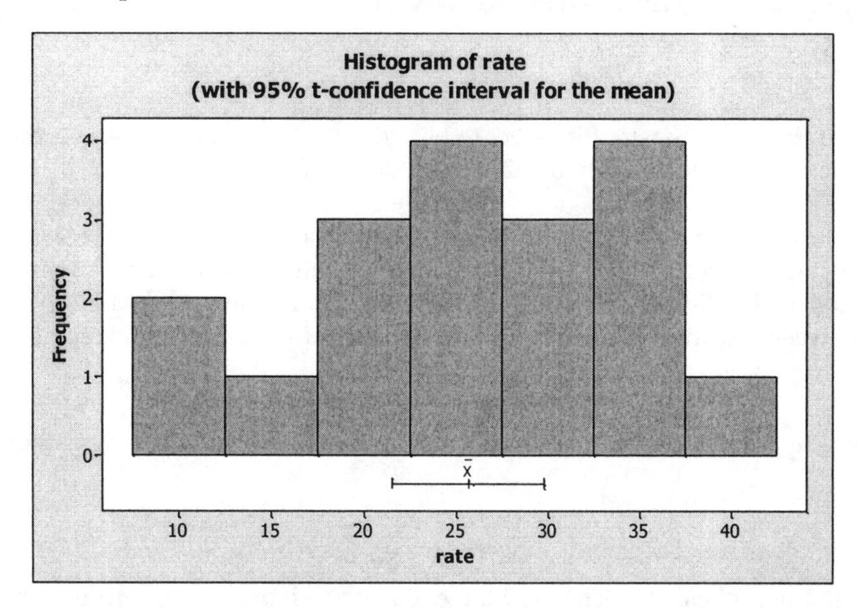

Since the sample size is equal to 18, $t_{\alpha/2}$ is the critical value from a *t*-distribution with 17 degrees of freedom. Select **Calc ➤ Probability Distributions ➤ t** from the menu to find the critical value you would use for a 95% confidence interval based on the $t(17)$ distribution.

You will need to click on Inverse cumulative probability. The Noncentrality parameter should be left at its default value of 0. To find the critical value $t_{\alpha/2}$ for a 95% confidence interval, the Input constant must be 0.975 as shown or 0.025.

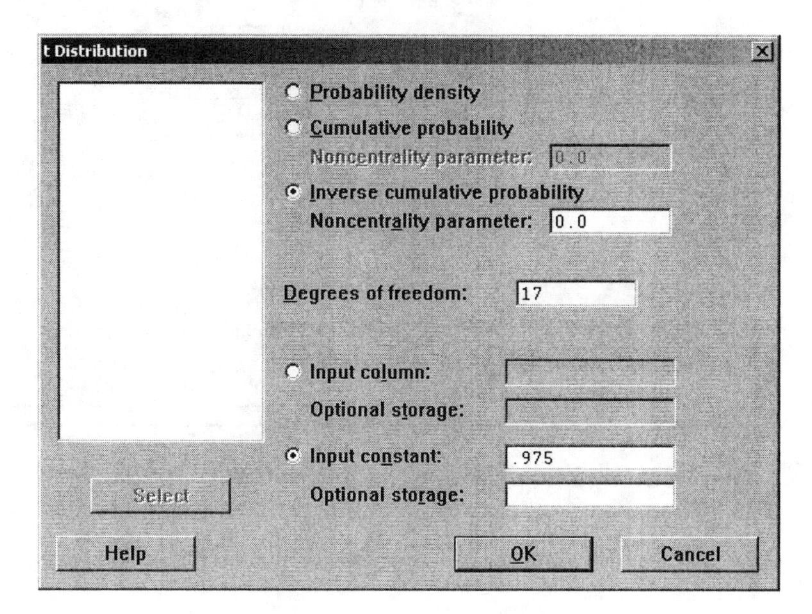

Inverse Cumulative Distribution Function

```
Student's t distribution with 17 DF

P( X <= x )        x
      0.975  2.10982
```

If the input constant is select to be $\alpha/2 = 0.025$ instead of $1 - \alpha/2$, the critical value will be calculated to be −2.10982 instead of 2.10982.

Example 15.2 of BPS tests whether or not a new cola recipe loses sweetness during storage. Trained tasters rate the sweetness before and after storage. Here and in EX15-02.MTW are the sweetness losses (sweetness before storage minus sweetness after storage) found by 10 tasters for one new cola recipe:

2.0 0.4 0.7 2.0 −0.4 2.2 −1.3 1.2 1.1 2.3

The null hypothesis is "no loss," and the alternative hypothesis says "there is a loss."

$$H_0: \mu = 0$$
$$H_a: \mu > 0$$

To do the hypothesis test choose **Stat ➤ Basic Statistics ➤ 1-Sample t** from the menu. As before, enter the column with the data into the variable box. In the Test mean box, enter 0.

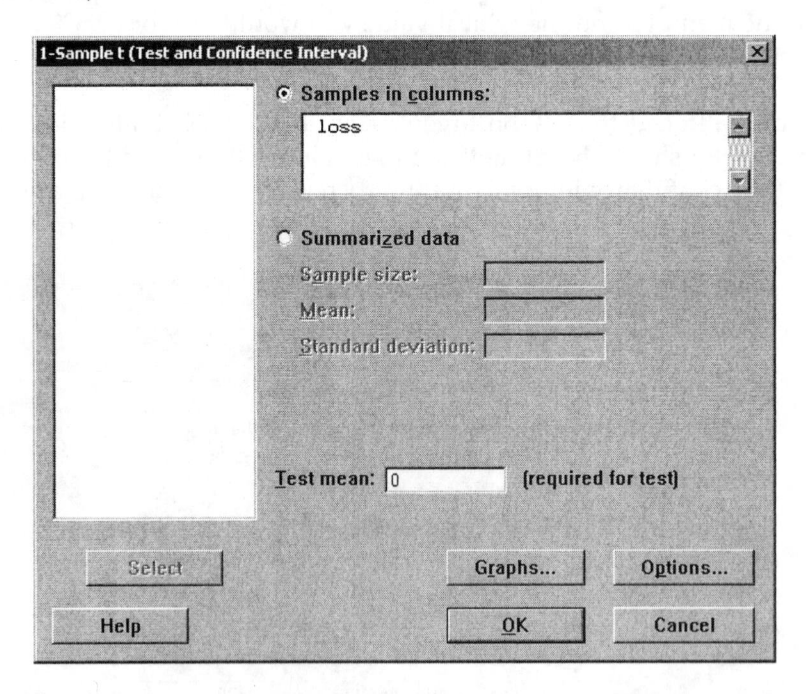

Click on the Options button and select the "less than" alternative.

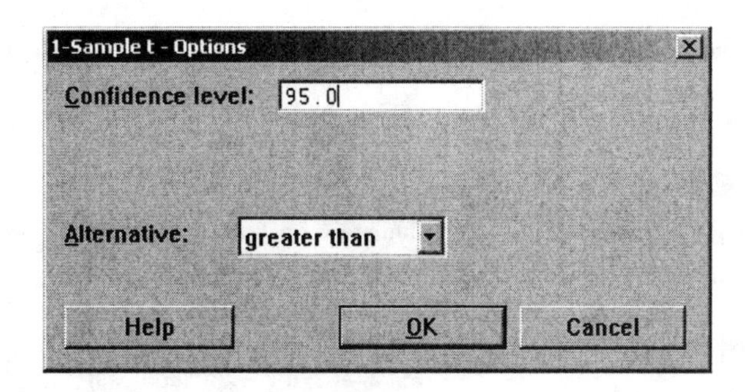

Minitab calculates the test statistic as

$$t = \frac{\bar{x} - \mu_0}{s/\sqrt{n}}$$

where $\bar{x}$ is the mean of the data, s is the sample standard deviation, n is the sample size, and μ_0 is the hypothesized population mean. As shown below, the test statistic t is calculated as 2.70. The P-value of this test is less than 0.012. There is strong evidence for a loss of sweetness. We can safely reject H_0.

One-Sample T: loss

```
Test of mu = 0 vs > 0


                                                95%
                                              Lower
Variable   N     Mean    StDev  SE Mean       Bound     T      P
loss      10  1.02000  1.19610  0.37824     0.32664  2.70  0.012
```

Note that when a one sided alternative is specified for the hypothesis test, a one-sided confidence interval is given by Minitab. You must select the "not equal" alternative to produce the confidence intervals described in BPS.

Matched Pairs

In a matched pairs study, subjects are matched in pairs and the outcomes are compared within each matched pair. In Example 16.3 of BPS, subjects worked a paper-and-pencil maze while wearing masks. Each mask was either unscented or carried a floral scent. The response variable is their mean time on three trials. Each subject worked the maze with both types of mask. Table 16.1 in BPS and TA16-01.MTW give the subject times. To assess whether the floral scent significantly improved performance, we test

$$H_0: \mu = 0$$
$$H_a: \mu > 0$$

where μ is the mean improvement if all executives received similar instruction.

To do a t test for matched pairs, select

Stat ➤ Basic Statistics ➤ Paired t

from the Minitab menu. Since Paired t evaluates the first sample minus the second sample, we select Unscented for the First sample and Scented for the Second sample.

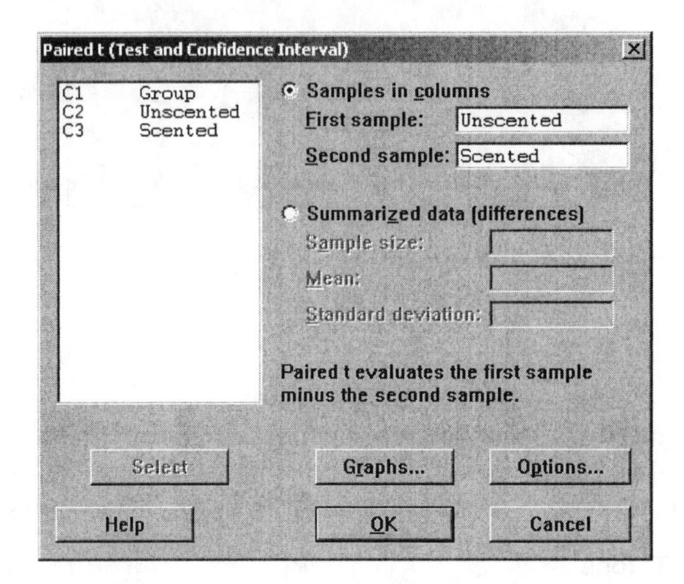

You must click on the Options button to specify that you are interested in improvements (time reductions) that are greater than 0.

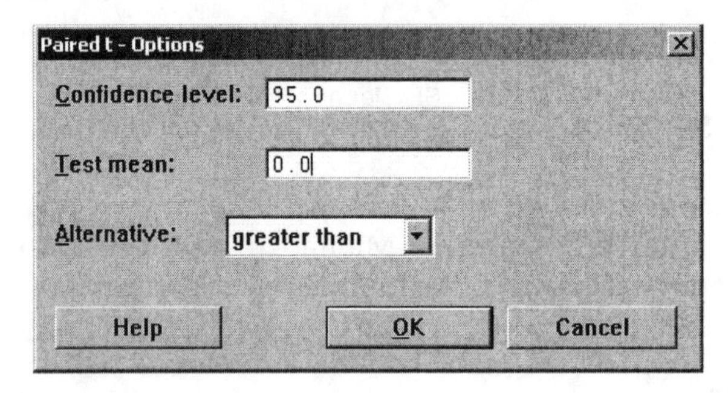

Paired T-Test and CI: Unscented, Scented

```
Paired T for Unscented - Scented

             N       Mean      StDev   SE Mean
Unscented   21    50.0143    14.3635    3.1344
Scented     21    49.0576    13.3856    2.9210
Difference  21   0.956667  12.547882  2.738172


95% lower bound for mean difference: -3.765909
T-Test of mean difference = 0 (vs > 0): T-Value = 0.35  P-Value = 0.365
```

The large value given for the P-value shows that the data do not support the claim that floral scents improve performance.

You can also compute a confidence interval for the mean improvement by selecting **Stat ➤ Basic Statistics ➤ Paired t**. In this case, you must select the "not equal" alternative in the Options subdialog box to avoid the one-sided interval.

Paired T-Test and CI: Unscented, Scented

```
Paired T for Unscented - Scented

             N       Mean      StDev   SE Mean
Unscented   21    50.0143    14.3635    3.1344
Scented     21    49.0576    13.3856    2.9210
Difference  21   0.956667  12.547882  2.738172


95% CI for mean difference: (-4.755061, 6.668394)
T-Test of mean difference = 0 (vs not = 0): T-Value = 0.35  P-Value = 0.730
```

EXERCISES

16.3 Select **Calc ➤ Probability Distributions ➤** *t* to find

(a) the value with probability 0.05 to its right under the $t(5)$ density curve.

(b) the value with probability 0.99 to its left under the $t(21)$ density curve.

16.4 You have an SRS of size 25 and calculate the one-sample *t* statistic. Select **Calc ➤ Probability Distributions ➤ t** to find the critical value t^* such that

(a) t has probability 0.025 to the right of t^*?

(b) t has probability 0.75 to the left of t^*?

16.5 Select **Calc ➤ Probability Distributions ➤ t** to find the critical value t^* you would use for a confidence interval for the mean of the population in each of the following situations?

(a) 95% confidence interval based on $n = 10$ observations

(b) 99% confidence interval from an SRS of 20 observations

(c) 80% confidence interval from a sample of size 7

16.6 Observational studies suggest that moderate use of alcohol reduces heart attacks, and that red wine may have special benefits. One reason may be that red wine contains polyphenols, substances that do good things to cholesterol in the blood and so may reduce the risk of heart attacks. In an experiment, healthy men were assigned at random to several groups. One group of 9 men drank half a bottle of red wine each day for two weeks. The level of polyphenols in their blood was measured before and after the two-week period. Here are the percent changes in level:

3.5 8.1 7.4 4.0 0.7 4.9 8.4 7.0 5.5

Select **Graph ➤ Stem-and-Leaf** from the menu to make a stemplot of the data. It is difficult to assess Normality from just nine observations. Select **Stat ➤ Basic Statistics ➤ 1-Sample t** from the menu to give a 90% *t* confidence interval for the mean percent change in blood polyphenols among all healthy men if all drank this amount of red wine.

16.7 The composition of the earth's atmosphere may have changed over time. To try to discover the nature of the atmosphere long ago, we can examine the gas in bubbles inside ancient amber. Amber is tree resin that has hardened and been trapped in rocks. The gas in bubbles within amber should be a sample of the atmosphere at the time the amber was formed. Measurements on specimens of amber from the late Cretaceous era (75 to 95 million years ago) give these percents of nitrogen:

63.4 65.0 64.4 63.3 54.8 64.5 60.8 49.1 51.0

Assume (this is not yet agreed on by experts) that these observations are an SRS from the late Cretaceous atmosphere.

(a) Graph the data, and comment on skewness and outliers. The t procedures will be only approximate for these data.

(b) Select **Stat ➤ Basic Statistics ➤ 1-Sample t** from the menu to give a 95% t confidence interval for the mean percent of nitrogen in ancient air.

16.10 The data of Exercise 16.7 suggest that the percent of nitrogen in the air during the Cretaceous era was quite different from the present 78.1%. Select **Stat ➤ Basic Statistics ➤ 1-Sample t** from the menu to carry out a test of

$$H_0: \mu = 78.1$$
$$H_a: \mu \neq 78.1$$

to assess the significance of the difference. (Give the test statistic, its P-value, and your conclusion.)

16.11 The design of controls and instruments affects how easily people can use them. A student project investigated this effect by asking 25 right-handed students to turn a knob (with their right hands) that moved an indicator by screw action. There were two identical instruments, one with a right-hand thread (the knob turns clockwise) and the other with a left-hand thread (the knob must be turned counterclockwise). Table 16.2 and TA 16-02.MTW give the times in seconds each subject took to move the indicator a fixed distance.

(a) Each of the 25 students used both instruments. Discuss briefly how you would use randomization in arranging the experiment.

(b) The project hoped to show that right-handed people find right-hand threads easier to use. What is the parameter μ for a matched pairs t test? State H_0 and H_a in terms of μ.

(c) Select **Stat ➤ Basic Statistics ➤ 1-Sample t** from the menu to carry out a test of your hypotheses. Give the P-value and report your conclusions.

16.12 Do "index funds" that simply buy and hold all the stocks in one of the stock market indexes, such as the Standard & Poor's 500-stock index, perform better than actively managed mutual funds? Compare the percent total return (price change plus dividends) of a large, actively managed fund with that of the Vanguard Index 500 fund for the 24 years 1977 to 2000. Vanguard did better by an average of 2.83% per year, and the standard deviation of the annual differences was 11.65%. Is there convincing evidence that the index fund does better?

(a) Describe in words the parameter μ for this comparison.

(b) State the hypotheses H_0 and H_a.

(c) Select **Stat ➤ Basic Statistics ➤ Paired t** from the menu to find the paired t statistic and its P-value. What do you conclude?

16.15 Exercise I.10 in BPS and EX0I-10.MTW gives data on the total amount of oil recovered from 64 oil wells in the same area. Take these wells to be an SRS of wells in this area.

(a) Select **Stat ➤ Basic Statistics ➤ 1-Sample t** to give a 95% t confidence interval for the mean amount of oil recovered from all wells in this area.

(b) The data are very skewed, with several high outliers. A new computer-intensive method that gives accurate confidence intervals without assuming any specific shape for the distribution gives the 95% confidence interval 40.28 to 60.32. How does the t interval compare with this?

16.16 A bank wonders whether omitting the annual credit card fee for customers who charge at least $2400 in a year will increase the amount charged on its credit cards. The bank makes this offer to an SRS of 200 of its credit card customers. It then compares how much these customers charge this year with the amount they charged last year. The mean increase in the sample is $332, and the standard deviation is $108.

(a) Is there significant evidence at the 1% level that the mean amount charged increases under the no-fee offer? State H_0 and H_a. Select **Stat ➤ Basic Statistics ➤ 1-Sample t** to carry out a t test.

(b) Select **Stat ➤ Basic Statistics ➤ 1-Sample t** to give a 99% confidence interval for the mean amount of the increase.

(c) The distribution of the amount charged is skewed to the right, but outliers are prevented by the credit limit that the bank enforces on each card. Use of the t procedures is justified in this case even though the population distribution is not Normal. Explain why.

16.18 A study of unexcused absenteeism among factory workers looked at a year's records for 668 workers in an English factory. The mean number of days absent was 9.88 and the standard deviation was 17.847 days. Regard these workers in this year as a random sample of all workers in all years as long as this factory does not change work conditions or worker benefits. Select **Stat ➤ Basic Statistics ➤ 1-Sample t** to see what you can say with 99% confidence about the mean number of unexcused absences for all workers.

16.20 A manufacturer of small appliances employs a market research firm to estimate retail sales of its products by gathering information from a sample of retail stores. This month an SRS of 75 stores in the Midwest sales region finds that these stores sold an average of 24 of the manufacturer's hand mixers, with standard deviation 11.

(a) Select **Stat ➤ Basic Statistics ➤ 1-Sample t** to give a 95% confidence interval for the mean number of mixers sold by all stores in the region.

(b) The distribution of sales is strongly right-skewed because there are many smaller stores and a few very large stores. The use of *t* in (a) is reasonably safe despite this violation of the Normality assumption. Why?

16.21 Here are measurements (in millimeters) of a critical dimension for 16 auto engine crankshafts:

224.120	224.001	224.017	223.982	223.989	223.961
223.960	224.089	223.987	223.976	223.902	223.980
224.098	224.057	223.913	223.999		

The dimension is supposed to be 224 mm and the variability of the manufacturing process is unknown. Is there evidence that the mean dimension is not 224 mm?

(a) Check the data graphically for outliers or strong skewness that might threaten the validity of the *t* procedures. What do you conclude?

(b) State H_0 and H_a and carry out a *t* test. Select **Stat ➤ Basic Statistics ➤ 1-Sample t** to find the *P*-value. What do you conclude?

16.22 To study the metabolism of insects, researchers fed cockroaches measured amounts of a sugar solution. After 2, 5, and 10 hours, they dissected some of the cockroaches and measured the amount of sugar in various tissues. Five roaches fed the sugar D-glucose and dissected after 10 hours had the following amounts (in micrograms) of D-glucose in their hindguts:

55.95 68.24 52.73 21.50 23.78

The researchers gave a 95% confidence interval for the mean amount of D-glucose in cockroach hindguts under these conditions. The insects are a random sample from a uniform population grown in the laboratory. We therefore expect responses to be Normal. Select **Stat ➤ Basic Statistics ➤ 1-Sample t** to find the confidence interval that the researchers gave.

16.23 How much do users pay for Internet service? Here are the monthly fees (in dollars) paid by a random sample of 50 users of commercial Internet service providers in August 2000:

20	40	22	22	21	21	20	10	20	20
20	13	18	50	20	18	15	8	22	25
22	10	20	22	22	21	15	23	30	12
9	20	40	22	29	19	15	20	20	20
20	15	19	21	14	22	21	35	20	22

(a) Select **Graph ➤ Stem-and-Leaf** from the menu to make a stemplot of the data. The data are not Normal: there are stacks of observations taking the same values, and the distribution is more spread out in both directions and somewhat skewed to the right. The t procedures are nonetheless approximately correct because $n = 50$ and there are no extreme outliers.

(b) Select **Stat ➤ Basics Statistics ➤ 1-Sample t** from the menu to give a 95% confidence interval for the mean monthly cost of Internet access in August 2000.

16.26 Hallux abducto valgus (call it HAV) is a deformation of the big toe that is not common in youth and often requires surgery. Doctors used X-rays to measure the angle (in degrees) of deformity in 38 consecutive patients under the age of 21 who came to a medical center for surgery to correct HAV. The angle is a measure of the seriousness of the deformity. Here are the data:

28	32	25	34	38	26	25	18	30	26	28	13	20
21	17	16	21	23	14	32	25	21	22	20	18	26
16	30	30	20	50	25	26	28	31	38	32	21	

It is reasonable to regard these patients as a random sample of young patients who require HAV surgery. Select **Stat ➤ Basics Statistics ➤ 1-Sample t** from the menu to give a 95% confidence interval for the mean HAV angle in the population of all such patients.

16.27 The data in the previous problem follow a normal distribution quite closely except for one patient with HAV angle 50 degrees, a high outlier.

(a) Select **Stat ➤ Basics Statistics ➤ 1-Sample t** from the menu to find the 95% confidence interval for the population mean based on the 37 patients who remain after you drop the outlier.

(b) Compare your interval in (a) with your interval from the previous problem. What is the most important effect of removing the outlier?

16.28 You are testing H_0: $\mu = 0$ against H_a: $\mu \neq 0$ based on an SRS of 20 observations from a normal population. Select **Calc ➤ Probability Distributions ➤ t** to find the values of the t statistic that are statistically significant at the $\alpha = 0.005$ level.

16.29 You have an SRS of 15 observations from a normally distributed population. Select **Calc ➤ Probability Distributions ➤ t** to find the critical value you would use to obtain a 98% confidence interval for the mean μ of the population.

16.30 Exercise 4.21 gives data on the annual returns (percent) for the Vanguard International Growth Fund and its benchmark index, the Morgan Stanley EAFE index. Does the fund significantly outperform its benchmark?

(a) Explain clearly why the matched pairs t test, not the two-sample t test, is the proper choice for answering this question.

(b) Select **Graph ➤ Stem-and-Leaf** to make a stemplot of the differences (fund − EAFE) for the 20 years. There is no reason to doubt approximate normality of the differences. (More detailed study shows that the differences follow a normal distribution quite closely.)

(c) Select **Stat ➤ Basic Statistics ➤ Paired t** to carry out the test and state your conclusion about the fund's performance.

16.31 Great white sharks are big and hungry. Here are the lengths in feet of 44 great whites:

18.7	12.3	18.6	16.4	15.7	18.3	14.6	15.8	14.9	17.6	12.1
16.4	16.7	17.8	16.2	12.6	17.8	13.8	12.2	15.2	14.7	12.4
13.2	15.8	14.3	16.6	9.4	18.2	13.2	13.6	15.3	16.1	13.5
19.1	16.2	22.8	16.8	13.6	13.2	15.7	19.7	18.7	13.2	16.8

(a) Examine these data for shape, center, spread, and outliers. The distribution is reasonably Normal except for one outlier in each direction. Because these are not extreme and preserve the symmetry of the distribution, use of the t procedures is safe with 44 observations.

(b) Select **Stat ➤ Basic Statistics ➤ 1-Sample t** to give a 95% confidence interval for the mean length of great white sharks. Based on this interval, is there significant evidence at the 5% level to reject the claim "Great white sharks average 20 feet in length"?

(c) It isn't clear exactly what parameter μ you estimated in (b). What information do you need to say what μ is?

16.32 In a randomized comparative experiment on the effect of calcium in the diet on blood pressure, researchers divided 54 healthy white males at random into two groups. One group received calcium; the other received a placebo. At the beginning of the study, the researchers measured many variables on the subjects. The paper reporting the study gives $\bar{x} = 114.9$ and $s = 9.3$ for the seated systolic blood pressure of the 27 members of the placebo group.

(a) Select **Stat ➤ Basic Statistics ➤ 1-Sample t** from the menu to give a 95% confidence interval for the mean blood pressure in the population from which the subjects were recruited.

(b) What assumptions about the population and the study design are required by the procedure you used in (a)? Which of these assumptions are important for the validity of the procedure in this case?

16.35 Differences of electric potential occur naturally from point to point on a body's skin. Is the natural electric field strength best for helping wounds to heal? If so, changing the field will slow healing. The research subjects are anesthetized newts. Make a razor cut in both hind limbs. Let one heal naturally (the control). Use an electrode to change the electric field in the other to half its Normal value. After two hours, measure the healing rate. Table 16.3 and TA16-03.MTW give the healing rates (in micrometers per hour) for 14 newts.

(a) As is usual, the paper did not report these raw data. Readers are expected to be able to interpret the summaries that the paper did report. The paper summarized the differences in the table as "−5.71 ± 2.82" and said, "All values are expressed as means ± standard error of the mean." Select **Stat ➤ Basic Statistics ➤ Display Descriptive Statistics** to show where the numbers −5.71 and 2.82 come from.

(b) The researchers want to know if changing the electric field reduces the mean healing rate for all newts. Select **Stat ➤ Basic Statistics ➤ Paired t** to carry out the test. State hypotheses, give the t statistic and its P-value, and give your conclusion. Is the result statistically significant at the 5% level? At the 1% level? (The researchers compared several field strengths and concluded that the natural strength is about right for fastest healing.)

16.37 Makers of generic drugs must show that they do not differ significantly from the "reference" drug that they imitate. One aspect in which drugs might differ is their extent of absorption in the blood. Table 7.5 and TA07-05.MTW give data taken from 20 healthy nonsmoking male subjects for one pair of drugs. This is a matched pairs design. Subjects 1 to 10 received the generic drug first, and Subjects 11 to 20 received the reference drug first. In all cases, a washout period separated the two drugs so that the first had disappeared from the blood before the subject took the second. The subject numbers in the table were assigned at random to decide the order of the drugs for each subject.

(a) Do a data analysis of the differences between the absorption measures for the generic and reference drugs. Is there any reason not to apply t procedures?

(b) Select **Stat ➤ Basics Statistics ➤ 1-Sample t** from the menu to answer the key question: Do the drugs differ significantly in absorption?

Chapter 17
Two-Sample Problems

Topics to be covered in this chapter:

Two-Sample *t* Procedures
The *F* Test for Equality of Variance

Two-Sample *t* Procedures

To perform a hypothesis test and compute a confidence interval of the difference between two population means, select

Stat ➤ Basic Statistics ➤ 2-Sample-t

from the menu. Check Samples in one column, Samples in different columns, or Summarized data depending on the format of your data.

Example 17.2 of BPS fits the two-sample setting. A researcher buried polyester strips in the soil to see how quickly they decay. Five of the strips, chosen at random, were dug up after two weeks. Another five were dug up after 16 weeks. The breaking strength (in pounds) of all 10 strips was measured and entered into EG17-02.MTW.

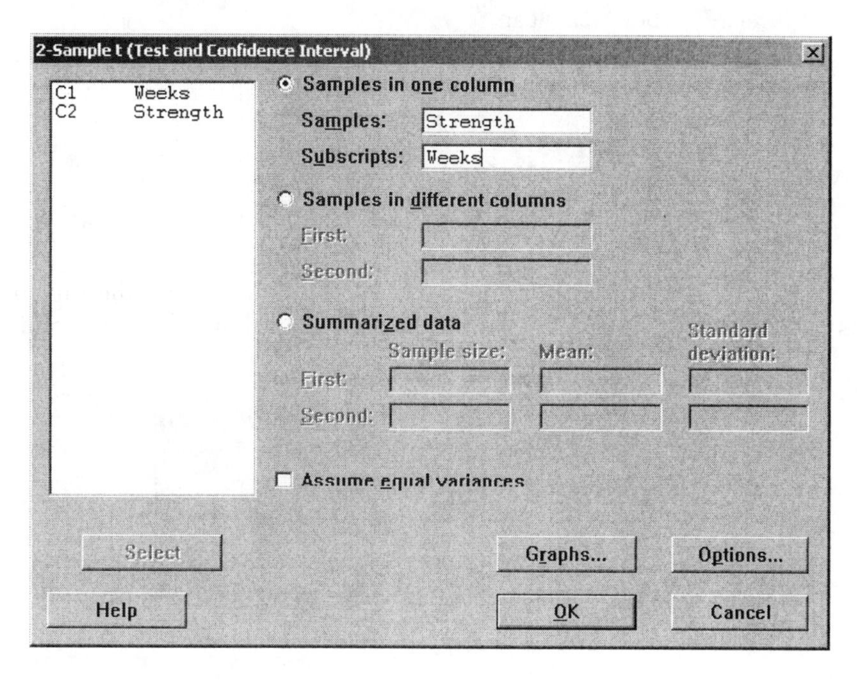

In EG17-02.MTW the samples are in one column, so we checked Samples in one column in the dialog box and entered the appropriate columns under Samples: and Subscripts.

We wish to test

$$H_0: \mu_2 = \mu_{16}$$
$$H_a: \mu_2 > \mu_{16}$$

so we click on the Options button and select the "greater than" alternative.

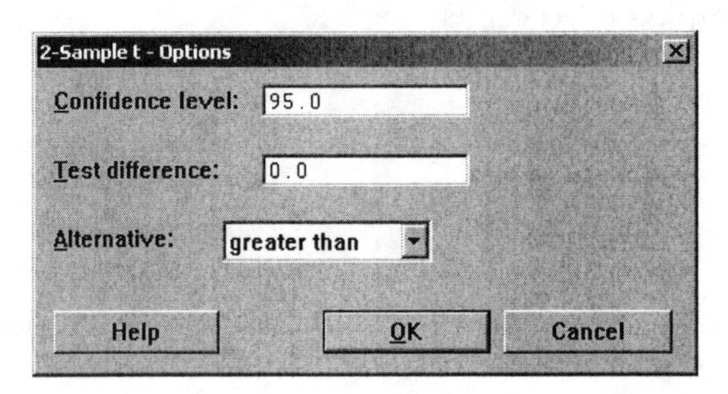

The following results show that the *P*-value was calculated to be 0.189. The experiment did not find convincing evidence that polyester decays more in 16 weeks than in two weeks. If a confidence interval is needed, the "not equal" alternative must be selected on the Options subdialog box.

Two-Sample T-Test and CI: Strength, Weeks

```
Two-sample T for Strength

Weeks  N    Mean  StDev  SE Mean
 2     5  123.80   4.60      2.1
16     5   116.4   16.1      7.2


Difference = mu ( 2) - mu (16)
Estimate for difference:  7.40000
95% lower bound for difference:  -8.55328
T-Test of difference = 0 (vs >): T-Value = 0.99  P-Value = 0.189  DF = 4
```

It is often helpful to obtain a graphical summary of the data in the two groups. Click on the Graphs button on the Two-Sample *t* dialog box and select Boxplots to produce side-by-side boxplots.

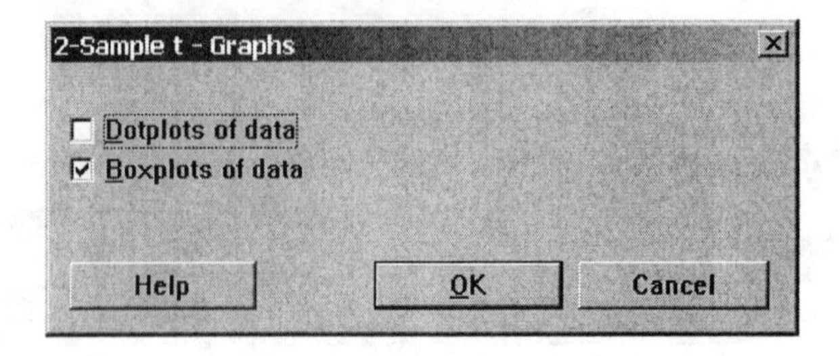

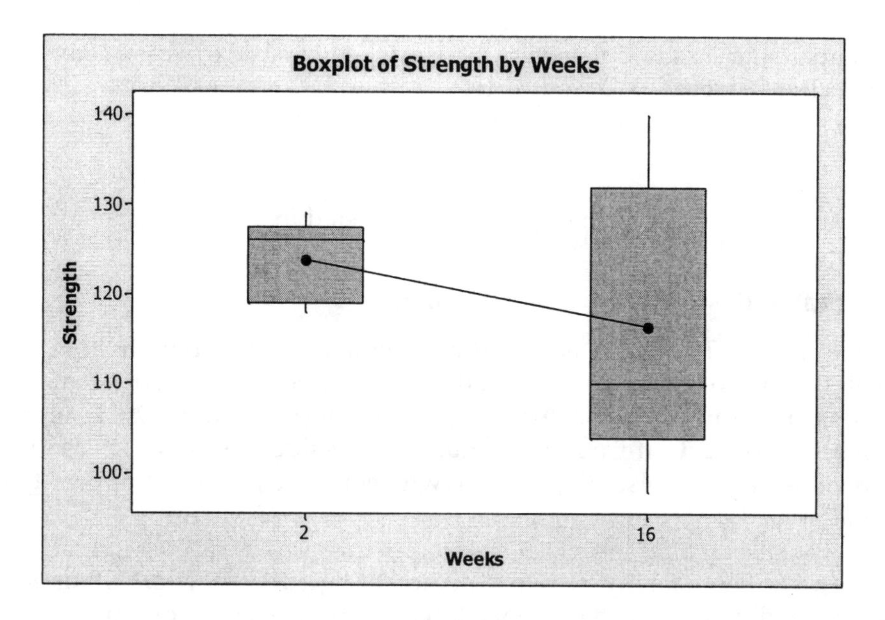

Unless the Assume equal variances box is checked, Minitab calculates the Two-Sample *t* test statistic as

$$t = \frac{\overline{X}_1 - \overline{X}_2}{\sqrt{\frac{s_1^2}{n_1} + \frac{s_2^2}{n_2}}}$$

This statistic has approximately a *t* distribution with degrees of freedom given by the Scatterthwaite approximation.

$$\text{df} = \frac{\left(\frac{s_1^2}{n_1} + \frac{s_2^2}{n_2}\right)^2}{\frac{1}{n_1 - 1}\left(\frac{s_1^2}{n_1}\right)^2 + \frac{1}{n_2 - 1}\left(\frac{s_2^2}{n_2}\right)^2}$$

Minitab truncates the number to an integer, if necessary.

If the Assume Equal Variances box is checked in the Two-Sample *t* dialog box, Minitab uses a pooled procedure, which assumes that the two populations have equal variances and "pools" the two sample variances to estimate the common population variance. The test statistic has a *t* distribution with exactly $n_1 + n_2 - 2$ degrees of freedom. The pooled procedure can be seriously in error if the variances are not equal. BPS recommends that the Assume Equal Variances box should never be checked.

The *F* Test for Equality of Variance

Example 17.2 describes an experiment to compare the mean breaking strengths of polyester fabric after being buried for 2 weeks and for 16 weeks. We might also compare the standard deviations to see whether strength loss is more or less variable after 16 weeks. We want to test

$$H_0 : \sigma_1 = \sigma_2$$
$$H_a : \sigma_1 \neq \sigma_2$$

The hypothesis of equal spread can be tested in Minitab using an F test by selecting

Stat ➤ Basic Statistics ➤ 2 Variances

from the menu. The F test is not recommended for distributions that are not normal. Before we calculate the F statistic, it is important to verify that the distributions are normal. This is done graphically by select **Graph ➤ Histogram** or **Graph ➤ Stem-and-Leaf** from the menu. Side-by-side boxplots such as those on the previous page are also useful to see whether the two groups appear to have the same spread.

To test the equality of two variances, select **Stat ➤ Basic Statistics ➤ 2 Variances** and fill out the dialog box. Check Samples in one column, Samples in different columns, or Summarized data depending on the format of your data.

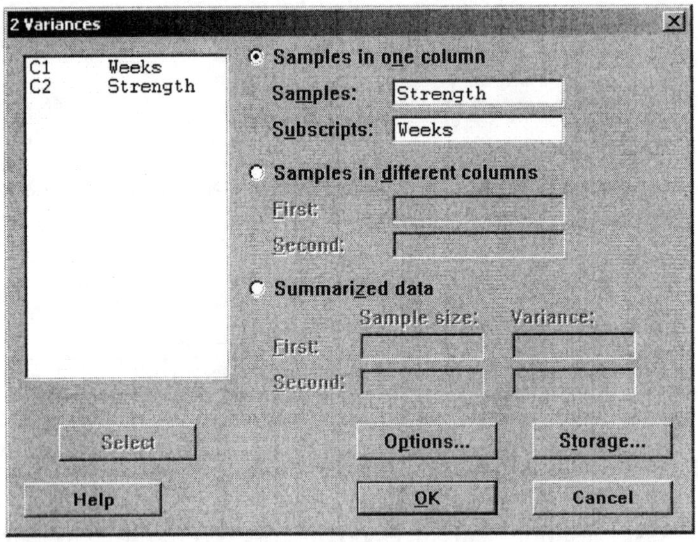

2 Variances
C1 Weeks
C2 Strength
Samples in one column
Samples: Strength
Subscripts: Weeks
Samples in different columns
First:
Second:
Summarized data
Sample size: Variance:
First:
Second:
Select
Options...
Storage...
Help
OK
Cancel

The F statistic is the ratio of the sample variances,

$$F = \frac{s_1^2}{s_2^2}$$

Minitab orders the groups alphabetically, so the ratio can be less than one. The P-value is equal to 0.033, so the difference between the spread on the two tests is statistically significant at the 5% level. The results from the test follow.

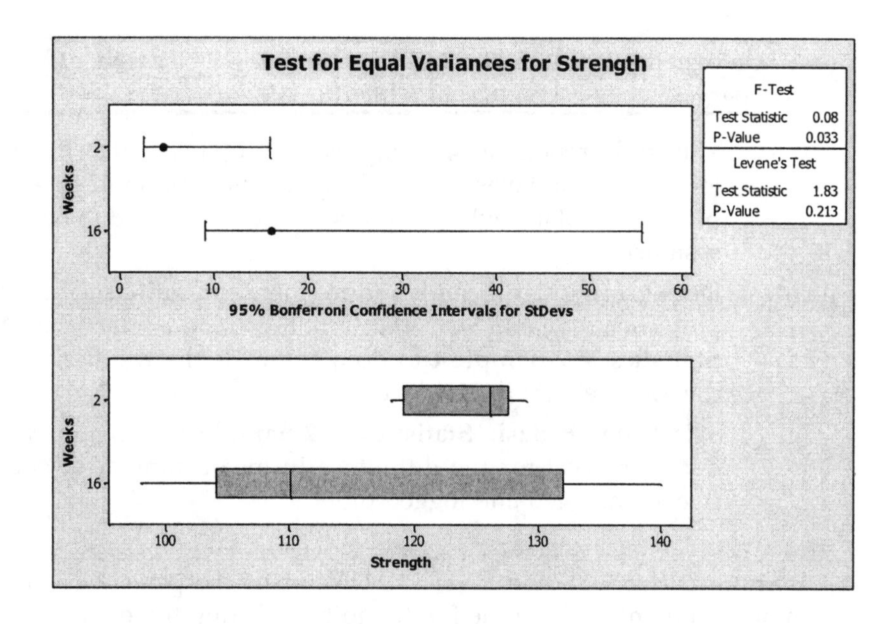

EXERCISES

17.7 Observational studies suggest that moderate use of alcohol reduces heart attacks, and that red wine may have special benefits. One reason may be that red wine contains polyphenols, substances that do good things to cholesterol in the blood and so may reduce the risk of heart attacks. In an experiment, healthy men were assigned at random to drink half a bottle of either red or white wine each day for two weeks. The level of polyphenols in their blood was measured before and after the two-week period. Here and in EX17-07.MTW are the percent changes in level for the subjects in both groups:

Red wine	3.5	8.1	7.4	4.0	0.7	4.9	8.4	7.0	5.5
White wine	3.1	0.5	–3.8	4.1	–0.6	2.7	1.9	–5.9	0.1

(a) Select **Stat ➤ Basic Statistics ➤ 2-Sample t** to see if there is good evidence that red wine drinkers gain more polyphenols on the average than white wine drinkers.

(b) Does this study give reason to think that it is drinking red wine, rather than some lurking variable, that causes the increase in blood polyphenols?

(c) Select **Stat ➤ Basic Statistics ➤ 2-Sample t** to give a 95% confidence interval for the difference (red minus white) in mean change in blood polyphenol levels.

17.8 "Conservationists have despaired over destruction of tropical rainforest by logging, clearing, and burning." These words begin a report on a statistical study of the effects of logging in Borneo. Here and in EX17-08.MTW are data on the number of tree species in 12 unlogged forest plots and 9 similar plots logged 8 years earlier:

Unlogged	22	18	22	20	15	21	13	13	19	13	19	15
Logged	17	4	18	14	18	15	15	10	12			

(a) The study report says, "Loggers were unaware that the effects of logging would be assessed." Why is this important? The study report also explains why the plots can be considered to be randomly assigned.

(b) Does logging significantly reduce the mean number of species in a plot after 8 years? State the hypotheses and select **Stat ➤ Basic Statistics ➤ 2-Sample t** to do a *t* test. Is the result significant at the 5% level? At the 1% level?

(c) Select **Stat ➤ Basic Statistics ➤ 2-Sample t** to give a 90% confidence interval for the difference in mean number of species between unlogged and logged plots

17.11 Exercise 2.19 in BPS and EX17-11.MTW gives the percent change in the mineral content of the spine for 47 mothers during three months of nursing a baby and for a control group of 22 women of similar age who were neither pregnant nor lactating.

(a) What two populations did the investigators want to compare? We must be willing to regard the women recruited for this observational study as SRSs from these populations.

(b) Make graphs to check for clear violations of the Normality condition. Use of the t procedures is safe for these distributions.

(c) Do these data give good evidence that on the average nursing mothers lose bone mineral? State the hypotheses. Select **Stat ➤ Basic Statistics ➤ 2-Sample t** to calculate the two-sample *t* test statistic, and give a *P*-value. What do you conclude?

17.13 In a randomized comparative experiment, researchers compared six white rats poisoned with DDT with a control group of six unpoisoned rats. Electrical measurements of nerve activity are the main clue to the nature of DDT poisoning. When a nerve is stimulated, its electrical response shows a sharp spike followed by a much smaller second spike. The experiment found that the second spike is larger in rats fed DDT than in normal rats. The researchers measured the height of the second spike as a percent of the first pike when a nerve in the rat's leg was stimulated. The results are here and in EX17-13.MTW:

Poisoned	12.207	16.869	25.050	22.429	8.456	20.589
Unpoisoned	11.074	9.686	12.064	9.351	8.182	6.642

Do poisoned rats differ significantly from unpoisoned rats in the study Select **Stat ➤ Basic Statistics ➤ 2-Sample t** to calculate the two-sample *t* test statistic, df, and *P*-value. What do you conclude?

17.20 Select **Stat ➤ Basic Statistics ➤ 2-Variances** to see if there is a statistically significant difference between the standard deviations of blood polyphenol level change in the red and white wine groups in Exercise 17.7.

17.21 Variation in species counts as well as mean counts may be of interest to ecologists. Select **Stat ➤ Basic Statistics ➤ 2-Variances** to see if the data in Exercise 17.8 give evidence that logging affects the variation in species counts among plots.

17.22 The sample variance for the treatment group in the DDT experiment of Exercise 17.13 is more than 10 times as large as the sample variance for the control group. Select **Stat ➤ Basic Statistics ➤ 2-Variances** to calculate the F statistic. Can you reject the hypothesis of equal population standard deviations at the 5% significance level? At the 1% level?

17.23 In March 2002 the Bureau of Labor Statistics contacted a large random sample of households for its Annual Demographic Supplement. Among the many questions asked were the education level and 2001 income of all persons aged 25 years or more in the household. The 24,042 whose education ended with a high school diploma had mean income $33,107.51 and standard deviation $30,698.36. There were 16,018 college graduates who had no higher degrees. Their mean income was $59,851.63 and the standard deviation was $57,001.32.

(a) It is clear without any calculations that the difference between the mean incomes for the two groups is significant at any practical level. Why?

(b) Select **Stat ➤ Basic Statistics ➤ 2-Sample t** to give a 99% confidence interval for the difference between the mean incomes in the populations of all adults with these levels of education.

(c) Income distributions are usually strongly skewed to the right. How does comparing $\bar{x}$ and s for these samples show the skewness? Why does the strong skewness not rule out using the two-sample t confidence interval?

17.24 "The phenomenon of road rage has been frequently discussed but infrequently examined." So begins a report based on interviews with randomly selected drivers. The respondents' answers to interview questions produced scores on an "angry/threatening driving scale" with values between 0 and 19. Here are summaries of the scores:

Group	n	$\bar{x}$	s
Male	596	1.78	2.79
Female	769	0.97	1.84

(a) We suspect that men are more susceptible to road rage than women. Select **Stat ➤ Basic Statistics ➤ 2-Sample t** to carry out a test of that hypothesis. (State hypotheses, find the test statistic and P-value, and state your conclusion.)

(b) The subjects were selected using random-digit dialing. The large sample sizes make the Normality condition unnecessary. There is one aspect of the data production that might reduce the validity of the data. What is it?

17.25 Healthy women aged 18 to 40 participated in a study of eating habits. Subjects were given bags of potato chips and bottled water and invited to snack freely. Interviews showed that some women were trying to restrain their diet out of concern about their weight. How much effect did these good intentions have on their eating habits? Here are the data on grams of potato chips consumed (note that the study report gave the standard error of the mean rather than the standard deviation):

Group	n	$\bar{x}$	SEM
Unrestrained	9	59	7
Restrained	11	32	10

Select **Stat ➤ Basic Statistics ➤ 2-Sample t** to give a 90% confidence interval that describes the effect of restraint. Based on this interval, is there a significant difference between the two groups? At what significance level does the interval allow this conclusion?

17.26 To study depression among adolescents, investigators administered the Children's Depression Inventory (CDI) to teenagers in rural Newfoundland, Canada. As is often the case in social science studies, there is some question about whether the subjects can be considered a random sample from an interesting population. We will ignore this issue. One finding was that "older adolescents scored significantly higher on the CDI." Higher scores indicate symptoms of depression. Here are summary data for two grades:

Group	n	$\bar{x}$	s
Grade 9	84	6.94	6.03
Grade 11	70	8.98	7.09

Select **Stat ➤ Basic Statistics ➤ 2-Sample t** to do an analysis to verify the quoted conclusion.

17.27 The study in the previous exercise also concluded that there were no sex differences in depression. Select **Stat ➤ Basic Statistics ➤ 2-Sample t** to verify this finding. Here are the summary data for males and females:

Group	n	$\bar{x}$	s
Males	112	7.38	6.95
Females	104	7.15	6.31

17.28 A "subliminal" message is below our threshold of awareness but may nonetheless influence us. Can subliminal messages help students learn math? A group of students who had failed the mathematics part of the City University of New York Skills Assessment Test agreed to participate in a study to find out. All received a daily subliminal message, flashed on a screen too rapidly to be consciously read. The treatment group of 10 students (chosen at random) was exposed to "Each day I am getting better in math." The control group of 8 students was exposed to a neutral message, "People are walking on the street." All students participated in a summer program designed to raise their math skills, and all took the assessment test again at the end of the program. Table 17.1 and TA17-01.MTW give data on the subjects' scores before and after the program.

(a) Is there good evidence that the treatment message brought about greater improvement in math scores than the neutral message? State the hypotheses. Select **Stat ➤ Basic Statistics ➤ 2-Sample t** to carry out a test, and state your conclusion. Is your result significant at the 5% level? At the 10% level?

(b) Select **Stat ➤ Basic Statistics ➤ 2-Sample t** to give a 90% confidence interval for the mean difference in gains between treatment and control.

17.37 A study of computer-assisted learning examined the learning of "Blissymbols" by children. Blissymbols are pictographs (think of Egyptian hieroglyphs) that are sometimes used to help learning-impaired children communicate. The researcher designed two computer lessons that taught the same content using the same examples. One lesson required the children to interact with the material, while in the other the children controlled only the pace of the lesson. Call these two styles "active" and "Passive." After the lesson, the computer presented a quiz that asked the children to identify 56 Blissymbols. Here and in EX17-37.MTW are the numbers of correct identifications by the 24 children in the active group:

29 28 24 31 15 24 27 23 20 22 23 21
24 35 21 24 44 28 17 21 21 20 28 16

The 24 children in the passive group had these counts of correct identifications:

16 14 17 15 26 17 12 25 21 20 18 21
20 16 18 15 26 15 13 17 21 19 15 12

(a) Is there good evidence that active learning is superior to passive learning? State the hypotheses. **Select Stat ➤ Basic Statistics ➤ 2-Sample t** to give a test and its *P*-value, and state your conclusion.

(b) What conditions does your test require? Which of these conditions can you use the data to check? Examine the data and report your results.

(c) **Select Stat ➤ Basic Statistics ➤ 2-Sample t** to give a 90% confidence interval for the difference in mean number of Blissymbols identified correctly by children after active and passive learning.

(d) **Select Stat ➤ Basic Statistics ➤ 1-Sample t** to give a 90% confidence interval for the mean number of Blissymbols identified correctly by children after the active computer lesson.

Chapter 18
Inference About a Population Proportion

Topics to be covered in this chapter:

Confidence Intervals for a Single Proportion
Significance Tests for a Proportion

Confidence Intervals for a Single Proportion

To compute a confidence interval and perform a hypothesis test of the proportion select

Stat ➤ Basic Statistics ➤ 1 Proportion

from the menu. In the dialog box, enter the column(s) containing the sample data. You can also enter summary data if they are in that form.

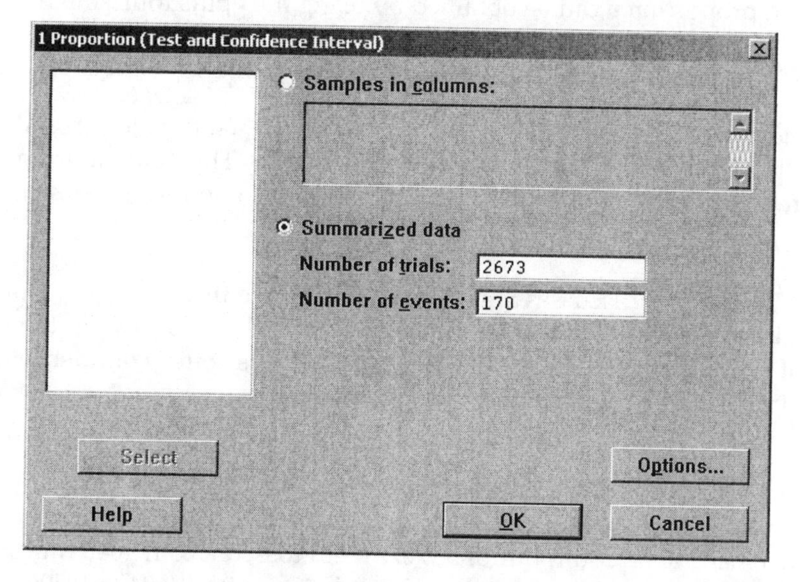

In Example 18.4 of BPS, a sample survey found that 170 of a sample of 2673 adult heterosexuals had multiple partners. The sample size is $n = 2673$ and the count of successes is $X = 170$. To make a 99% confidence interval for the proportion p of all adult heterosexuals with multiple partners, select **Stat ➤ Basic Statistics ➤ 1 Proportion** from the menu and fill in the summarized data as shown in the dialog box. Click on the Options button and set the Confidence level to 99 and make sure the "not equal" alternative is selected. If you check the box specifying that Minitab use tests and intervals based on the normal distribution, the calculations will be the same as those in BPS.

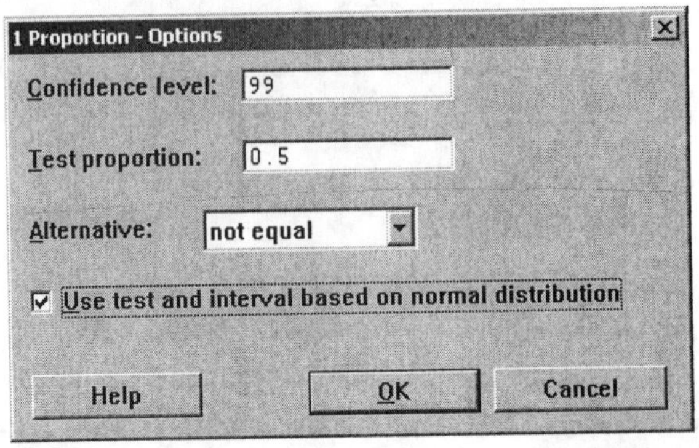

1 Proportion - Options

Confidence level: 99

Test proportion: 0.5

Alternative: not equal

☑ Use test and interval based on normal distribution

Help OK Cancel

Test and CI for One Proportion

```
Test of p = 0.5 vs p not = 0.5

Sample     X     N  Sample p            99% CI             Z-Value  P-Value
1        170  2673  0.063599  (0.051441, 0.075757)   -45.12    0.000
```

As described in Chapter 18 of BPS, a more accurate confidence interval for the population proportion p can be obtained by using the "plus four" method.

$$\tilde{p} = \frac{X+2}{n+4}$$

We can add two successes and two failures to the actual data to force Minitab to use the "plus four" method for confidence intervals. The "plus four" method is always recommended and is particularly important to do if samples sizes are not large.

Example 18.6 of BPS discusses Shaquille O'Neal's free throws. After working on his technique between seasons, he made 26 out of 39 free throws in the first two games of the following season. To compute a "plus four" confidence interval, enter 43 trials and 28 events in the 1 Proportion dialog box.

Significance Tests for a Proportion

The French naturalist Count Buffon (1707–1788) tossed a coin 4040 times. He got 2048 heads. The sample proportion of heads is $\hat{p} = 2048/4040 = 0.5069$. Is this evidence that Buffon's coin was not balanced? To answer this question we test

$$H_0 : p = 0.5$$
$$H_a : p \neq 0.5$$

Click on the Options button on the 1 Proportion dialog box to open the Options subdialog box. The Options subdialog box allows you to specify the test proportion and alternative hypothesis, and to check a box specifying that Minitab use tests and intervals based on the normal distribution.

By default, Minitab uses an exact method to calculate the test probability. If you choose to use a normal approximation, Minitab calculates the test statistic (z) as

$$z = \frac{\hat{p} - p_0}{\sqrt{\frac{p_0(1-p_0)}{n}}}$$

where $\hat{p}$ is the observed probability equal to X/n, X is the observed number of successes in n trials, and p_0 is the hypothesized probability. The probabilities are obtained from a standard normal distribution table.

1 Proportion - Options

Confidence level: 99

Test proportion: 0.5

Alternative: not equal

☑ Use test and interval based on normal distribution

Help | OK | Cancel

Test and CI for One Proportion

```
Test of p = 0.5 vs p not = 0.5

Sample     X     N  Sample p          99% CI             Z-Value  P-Value
1       2048  4040  0.506931  (0.486670, 0.527191)     0.88     0.378
```

Minitab gives the *P*-values as 0.378. This means that a proportion of heads as far from one-half as Buffon's would happen more than 30% of the time when a balanced coin is tossed 4040 times. Buffon's result doesn't show that his coin is unbalanced.

EXERCISES

18.9 A random sample of students who took the college entrance SAT twice found that 427 of the respondents had paid for coaching courses and that the remaining 2733 had not. Select **Stat ➤ Basic Statistics ➤ 1 Proportion** to give a 99% confidence interval for the proportion of coaching among students who retake the SAT.

18.11 The *New York Times* and CBS News conducted a nationwide survey of 1048 randomly selected 13- to 17-year-olds. Of these teenagers, 692 had a television in their room.

(a) Check that we can use the large-sample confidence interval.

(b) Select **Stat ➤ Basic Statistics ➤ 1 Proportion** to give a 95% confidence interval for the proportion of all teens who have a TV set in their room.

(c) The news article says, "In theory, in 19 cases out of 20, the survey results will differ by no more than three percentage points in either direction from what would have been obtained by seeking out all American teenagers." Explain how your results agree with this statement.

18.14 The elderly fear crime more than younger people, even though they are less likely to be victims of crime. One of the few studies that looked at older blacks recruited a random sample of 56 black women over the age of 65 from Atlantic City, New Jersey. Of these women, 27 said that they "felt vulnerable" to crime.

(a) Give the two estimates $\hat{p}$ and $\tilde{p}$ of the proportion p of all elderly black women in Atlantic City who feel vulnerable to crime. There is little difference between them. This is generally true when $\hat{p}$ is not close to either 0 or 1.

(b) Select **Stat ➤ Basic Statistics ➤ 1 Proportion** to give both the large-sample 95% confidence interval and the "plus four" 95% confidence interval for p. The "plus four" interval is a bit narrower. This is generally true when $\hat{p}$ is not close to either 0 or 1.

18.15 Social scientists rarely ask about religious behaviors such as prayer. Investigators asked 127 undergraduate students "from courses in psychology and communications" about prayer and found that 107 prayed at least a few times a year.

(a) Select **Stat ➤ Basic Statistics ➤ 1 Proportion** to give the "plus four" 99% confidence interval for the proportion p of all students who pray.

(b) To use any inference procedure, we must be willing to regard these 127 students, as far as their religious behavior goes, as an SRS from the population of all undergraduate students. Do you think it is reasonable to do this? Why or why not?

18.18 Spinning a coin, unlike tossing it, may not give heads and tails equal probabilities. I spun a penny 200 times and got 83 heads. Select **Stat ➤ Basic Statistics ➤ 1 Proportion** to see how significant this evidence is against equal probabilities. (State hypotheses, give the test statistic and its P-value, and state your conclusion.)

18.19 A random sample of 1048 13- to 17-year-olds found that 692 had a television set in their room. Select **Stat ➤ Basic Statistics ➤ 1 Proportion** to see if this is good evidence that more than half of all teens have a TV in their room. (State hypotheses, give the test statistic and its P-value, and state your conclusion.)

18.21 A random sample of 1318 Internet users was asked where they will go for information the next time they need information about health or medicine; 606 said that they would use the Internet. Select **Stat ➤ Basic Statistics ➤ 1 Proportion** to give a 99% confidence interval for the proportion of all Internet users who feel this way. Be sure to check that the conditions for use of your method are met.

18.22 The proportion of drivers who use seat belts depends on things like age, gender, ethnicity, and local law. As part of a broader study, investigators observed a random sample of 117 female Hispanic drivers in Boston; 68 of these drivers were wearing seat belts. Select **Stat ➤ Basic Statistics ➤ 1 Proportion** to five a 95% confidence interval for the proportion of all female Hispanic drivers in Boston who wear seat belts. Be sure to check that the conditions for use of your method are met.

18.23 A Gallup Poll on energy use asked 512 randomly selected adults if they favored "increasing the use of nuclear power as a major source of energy." Gallup reported that 225 said "Yes." Does this poll give good evidence that less than half of all adults favor increased use of nuclear power? Select **Stat ➤ Basic Statistics ➤ 1 Proportion** to see. State hypotheses, give the test statistic and its *P*-value, and state your conclusion.

18.24 Do the data in Exercise 18.22 give good reason to conclude that more than half of Hispanic female drivers in Boston wear seat belts? Select **Stat ➤ Basic Statistics ➤ 1 Proportion** to see. State hypotheses, give the test statistic and its *P*-value, and state your conclusion.

18.25 The College Alcohol Study interviewed an SRS of 14,941 college students about their drinking habits. The sample was stratified using 140 colleges as strata, but the overall effect is close to an SRS of students. The response rate was between 60% and 70% at most colleges. This is quite good for a national sample, though nonresponse is as usual the biggest weakness of this survey. Of the students in the sample, 10,010 supported cracking down on underage drinking. Select **Stat ➤ Basic Statistics ➤ 1 Proportion** to give a 99% confidence interval for the proportion of all college students who feel this way.

18.26 A random-digit-dialing telephone survey of 880 drivers asked, "Recalling the last ten traffic lights you drove through, how many of them were red when you entered the intersections?" Of the 880 respondents, 171 admitted that at least one light had been red.

(a) Select **Stat ➤ Basic Statistics ➤ 1 Proportion** to give a 95% confidence interval for the proportion of all drivers who ran one or more of the last ten red lights they met.

(b) Nonresponse is a practical problem for this survey—only 21.6% of calls that reached a live person were completed. Another practical problem is that people may not give truthful answers. What is the likely direction of the bias: Do you think more or fewer than 171 of the 880 respondents really ran a red light? Why?

18.27 Most soybeans grown in the United States are genetically modified to, for example, resist pests and so reduce use of pesticides. Because some nations do not accept genetically modified (GM) foods, grain-handling facilities routinely test soybean shipments for the presence of GM beans. In a study of the accuracy of these tests, researchers submitted lots of soybeans containing 1% GM beans to 23 randomly selected facilities. Eighteen detected the GM beans.

(a) Show that the conditions for the large-sample confidence interval are not met. Show that the conditions for the "plus four" interval are met.

(b) Select **Stat ➤ Basic Statistics ➤ 1 Proportion** to use the "plus four" method to give a 90% confidence interval for the percent of all grain-handling facilities that will correctly detect 1% of GM beans in a shipment.

18.30 The National AIDS Behavioral Surveys (Example 18.1) also interviewed a sample of adults in the cities where AIDS is most common. This sample included 803 heterosexuals who reported having more than one sexual partner in the past year. We can consider this an SRS of size 803 from the population of all heterosexuals in high-risk cities who have multiple partners. These people risk infection with the AIDS virus. Yet 304 of the respondents said they never use condoms. Select **Stat ➤ Basic Statistics ➤ 1 Proportion** to see if this is strong evidence that more than one-third of this population never use condoms.

18.31 The Gallup Poll asked a sample of 1785 adults, "Did you, yourself, happen to attend church or synagogue in the last 7 days?" Of the respondents, 750 said "Yes." Treat Gallup's sample as an SRS of all American adults. Select **Stat ➤ Basic Statistics ➤ 1 Proportion** to give a 99% confidence interval for the proportion of all adults who claimed that they attended church or synagogue during the week preceding the poll. (The proportion who actually attended is no doubt lower—some people say "Yes" if they usually attend, or often attend, or sometimes attend.)

18.32 Do the results of the poll in Exercise 18.31 provide good evidence that fewer than half of the population would claim to have attended church or synagogue? Select **Stat ➤ Basic Statistics ➤ 1 Proportion** to see.

18.35 One-sample procedures for proportions, like those for means, are used to analyze data from matched pairs designs. Here is an example. Each of 50 subjects tastes two unmarked cups of coffee and says which he or she prefers. One cup in each pair contains instant coffee, the other fresh-brewed coffee. Thirty-one of the subjects prefer the fresh-brewed coffee.

Take p to be the proportion of the population who would prefer fresh-brewed coffee in a blind tasting.

(a) Select **Stat ➤ Basic Statistics ➤ 1 Proportion** to test the claim that a majority of people prefer the taste of fresh-brewed coffee. State hypotheses and report the z statistic and its P-value. Is your result significant at the 5% level? What is your practical conclusion?

(b) Select **Stat ➤ Basic Statistics ➤ 1 Proportion** to find a 90% confidence interval for p.

(c) When you do an experiment like this, in what order should you present the two cups of coffee to the subjects?

18.38 A nationwide random survey of 1500 adults asked about attitudes toward "alternative medicine" such as acupuncture, massage therapy, and herbal therapy. Among the respondents, 660 said they would use alternative medicine if traditional medicine was not producing the results they wanted.

(a) Select **Stat ➤ Basic Statistics ➤ 1 Proportion** to give a 95% confidence interval for the proportion of all adults who would use alternative medicine.

(b) Write a short paragraph for a news report based on the survey results.

Chapter 19
Comparing Two Proportions

Topics to be covered in this chapter:

Confidence Interval for Comparing Proportions
More Accurate Confidence Intervals
Significance Tests for Comparing Proportions

Confidence Interval for Comparing Proportions

To compute a confidence interval and perform a hypothesis test of the difference between two proportions select

Stat ➤ Basic Statistics ➤ 2 Proportions

from the menu. You can enter the two samples into a single column with subscripts in a second column, or you can enter the samples into two different columns. You can also use summary data if they are in that form. The Options subdialog box allows you to specify the confidence level, the null hypothesis value, and the form of the alternative hypothesis. You can also specify that Minitab use a pooled estimate of p for the hypothesis test.

Example 19.1 of BPS Example 19.1 describes a study of the effect of preschool on later use of social services. One group of 62 attended preschool as 3- and 4-year-olds. A control group of 61 children from the same area did not attend preschool. Thus the sample sizes are $n_1 = 61$ and $n_2 = 62$. In the past ten years, 38 of the preschool sample and 49 of the control sample have needed social services (mainly welfare). Here is the data summary:

Population	n	X	$\hat{p} = X/n$
1 (control)	61	49	0.803
2 (preschool)	62	38	0.613

The difference $p_1 - p_2$ measures the effect of preschool in reducing the proportion of people who later need social services. To compute a 95% confidence interval for $p_1 - p_2$, select **Stat ➤ Basic Statistics ➤ 2 Proportions** from the Minitab menu. Enter the summarized data in the dialog box and click on the Options button.

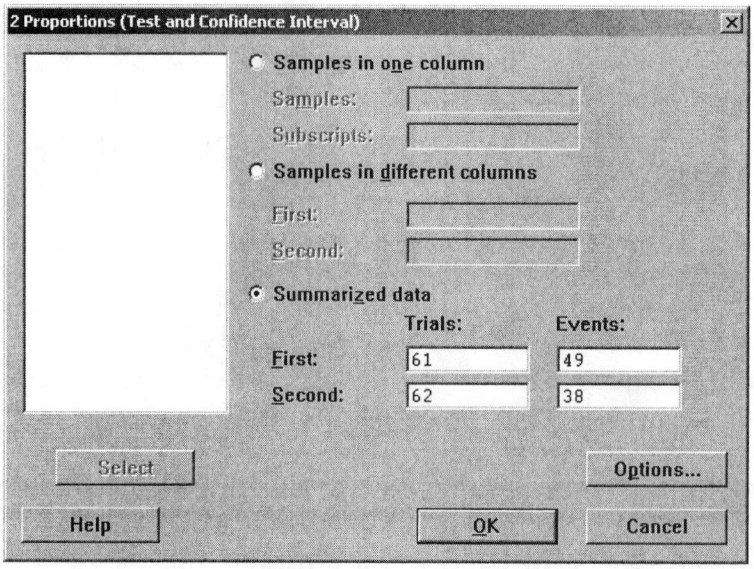

2 Proportions (Test and Confidence Interval)

Samples in one column
Samples:
Subscripts:
Samples in different columns
First:
Second:
Summarized data
Trials: Events:
First: 61 49
Second: 62 38
Select
Options...
Help
OK
Cancel

In the Options subdialog box, make sure that the Confidence level is set to 95 and the Alternative is "not equal". Minitab calculates the confidence interval as

$$\hat{p}_1 - \hat{p}_2 \pm z^* \sqrt{\frac{\hat{p}_1(1-\hat{p}_1)}{n_1} + \frac{\hat{p}_2(1-\hat{p}_2)}{n_2}},$$

where $\hat{p}_1$ and $\hat{p}_2$ are the observed probabilities of sample one and sample two, respectively, and $\hat{p} = X/n$, where X is the observed success in n trials.

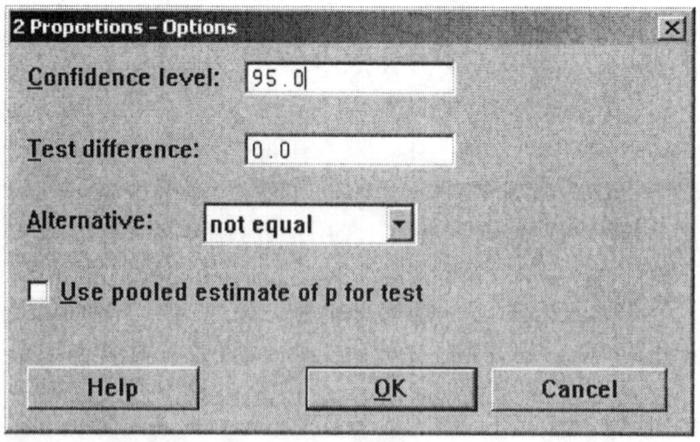

2 Proportions - Options

Confidence level: 95.0
Test difference: 0.0
Alternative: not equal
Use pooled estimate of p for test
Help
OK
Cancel

The following results show that we are 95% confident that the percent needing social services is somewhere between 3.3 and 34.7 percentage points lower among people who attended preschool. The confidence interval is wide because the samples are quite small.

Test and CI for Two Proportions

```
Sample    X   N  Sample p
1        49  61  0.803279
2        38  62  0.612903


Difference = p (1) - p (2)
Estimate for difference:  0.190375
95% CI for difference:  (0.0333680, 0.347383)
Test for difference = 0 (vs not = 0):  Z = 2.38  P-Value = 0.017
```

More Accurate Confidence Intervals

A simple modification improves the accuracy of confidence interval for comparing propositions. As with a single proportion, the interval is called the "plus four" interval because you add four imaginary observations, one success and one failure in each of the two samples. That is, we let

$$\tilde{p}_1 = \frac{X_1 + 1}{n_1 + 2} \text{ and } \tilde{p}_2 = \frac{X_2 + 1}{n_2 + 2}.$$

Population	n	X	$\hat{p} = X/n$
1 (control)	$61 + 2 = 63$	$49 + 1 = 50$	0.794
2 (preschool)	$62 + 2 = 64$	$38 + 1 = 39$	0.609

To compute the new interval, select **Stat ➤ Basic Statistics ➤ 2 Proportions** from the menu, enter the new summarized data, and click OK.

Test and CI for Two Proportions

```
Sample    X   N  Sample p
1        50  63  0.793651
2        39  64  0.609375


Difference = p (1) - p (2)
Estimate for difference:  0.184276
95% CI for difference:  (0.0284760, 0.340076)
Test for difference = 0 (vs not = 0):  Z = 2.32  P-Value = 0.020
```

The "plus four" interval is not very different from the large-sample interval in this case because the sample sizes are large. The "plus four" interval is generally much more accurate than the large-sample interval when the samples are small.

Significance Tests for Comparing Proportions

We also can use Minitab to do a significance tests to help us decide if the effect we see in the samples is really there in the populations. The null hypothesis says that there is no difference between the two populations: $H_0 : p_1 = p_2$.

In Example 19.4 in BPS, researchers ask the question "Would you marry a person from a lower social class than your own?" Of the 149 men in the sample, 91 said "Yes." Among the 236 women, 117 said "Yes." To see if there is a statistically significant difference we have a two-sided alternative:

$$H_0 : p_1 = p_2$$
$$H_a : p_1 \neq p_2$$

Select **Stat ➤ Basics Statistics ➤ 2 Proportions** to perform the significance test. Enter the summary information in the dialog box and click on the Options button.

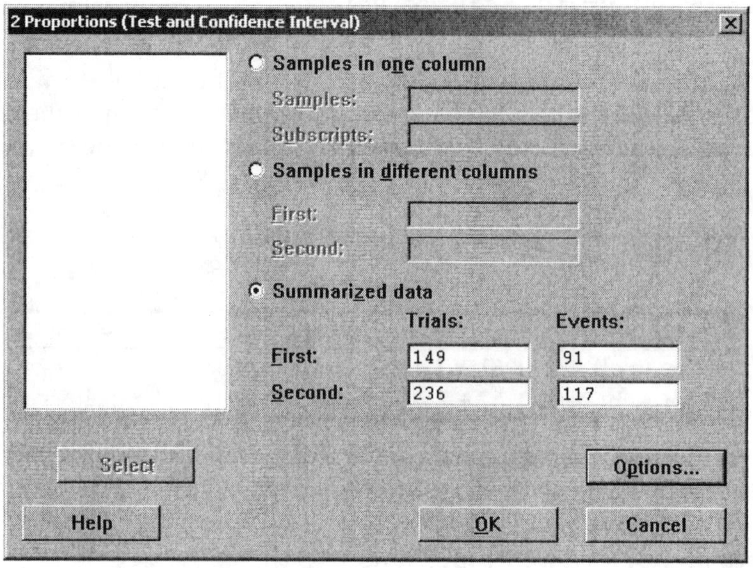

2 Proportions (Test and Confidence Interval)

Samples in one column
Samples:
Subscripts:
Samples in different columns
First:
Second:
Summarized data

	Trials:	Events:
First:	149	91
Second:	236	117

Select | Options... | Help | OK | Cancel

In the subdialog box, make sure that Test difference is equal to 0, the Alternative is "not equal" and the Use pooled estimate of p for test is selected so that Minitab uses a pooled estimate of p for the hypothesis test and calculates z as

$$z = \frac{\hat{p}_1 - \hat{p}_2}{\sqrt{\hat{p}(1-\hat{p})\left(\frac{1}{n_1} + \frac{1}{n_2}\right)}}, \text{ where } \hat{p} = \frac{X_1 + X_2}{n_1 + n_2}.$$

If Use pooled estimate of p for test is not checked, Minitab uses separate estimates of p for each population.

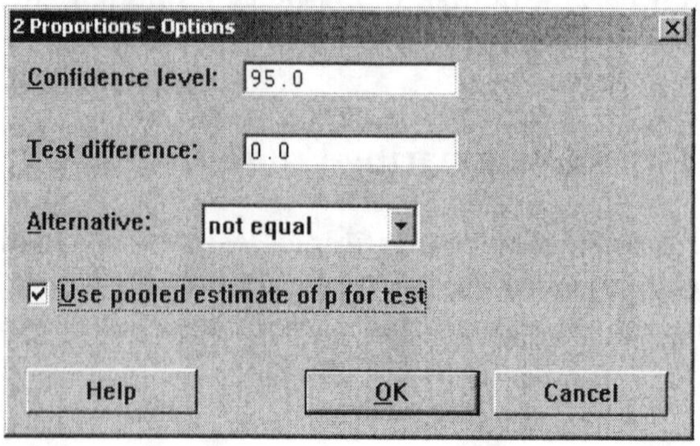

2 Proportions - Options

Confidence level: 95.0
Test difference: 0.0
Alternative: not equal
☑ Use pooled estimate of p for test

Help | OK | Cancel

Because the *P*-value = 0.027, the results are statistically significant at the $\alpha = 0.05$ level. There is good evidence that men are more likely than women to say they will marry someone from a lower social class.

Test and CI for Two Proportions

```
Sample    X    N  Sample p
1        91  149  0.610738
2       117  236  0.495763


Difference = p (1) - p (2)
Estimate for difference:  0.114976
95% CI for difference:  (0.0139890, 0.215962)
Test for difference = 0 (vs not = 0):  Z = 2.20  P-Value = 0.027
```

As described in Chapter 18 of BPS, a more accurate confidence interval for the difference in population proportions can be obtained by using the "plus four" estimates.

EXERCISES

19.1 The elderly fear crime more than younger people, even though they are less likely to be victims of crime. One of the few studies that looked at older blacks recruited random samples of 56 black women and 63 black men over the age of 65 from Atlantic City, New Jersey. Of the women, 27 said they "felt vulnerable" to crime; 46 of the men said this.

(a) What proportion of women in the sample feel vulnerable? Of men? Men are victims of crime more often than women, so we expect a higher proportion of men to feel vulnerable.

(b) Select **Stat ➤ Basic Statistics ➤ 2 Proportions** to give a 95% confidence interval for the difference (men minus women).

19.2 Nicotine patches are often used to help smokers quit. Does giving medicine to fight depression help smokers quit? A randomized double-blind experiment assigned nicotine patches to 244 smokers who wanted to stop and both a patch and the antidepressant drug bupropion to another 245 smokers. Results: After a year, 40 subjects in the nicotine patch group still abstained from smoking, as did 87 in the patch-plus-drug group. Select **Stat ➤ Basic Statistics ➤ 2 Proportions** to give a 99% confidence interval for the difference (treatment minus control) in the proportion of smokers who quit.

19.3 In 2002 the Supreme Court ruled that schools could require random drug tests of students participating in competitive after-school activities such as athletics. Does drug testing reduce use of illegal drugs? A study compared two similar high schools in Oregon. Wahtonka High School tested athletes at random and Warrenton High School did not. In a confidential survey, 7 of 135 athletes at Wahtonka and 27 of 141 athletes at Warrenton said they were using drugs. Regard these athletes as SRSs from the populations of athletes at similar schools with and without drug testing.

(a) You should not use the large-sample confidence interval. Why not?

(b) The "plus four" method adds two observations, a success and a failure, to each sample. What are the sample sizes and the counts of drug users after you do this?

(c) Select **Stat ➤ Basic Statistics ➤ 2 Proportions** to give the "plus four" 95% confidence interval for the difference between the proportion of athletes using drugs at schools with and without testing.

19.4 A study of injuries to in-line skaters used data from the National Electronic Injury Surveillance System, which collects data from a random sample of hospital emergency rooms. The researchers interviewed 161 people who came to emergency rooms with injuries from in-line skating. Wrist injuries (mostly fractures) were the most common.

(a) The interviews found that 53 people were wearing wrist guards and 6 of these had wrist injuries. Of the 108 who did not wear wrist guards, 45 had wrist injuries. Why should we not use the large-sample confidence interval for these data?

(b) Select **Stat ➤ Basic Statistics ➤ 2 Proportions** to give the "plus four" 95% confidence interval for the difference between the two population proportions of wrist injuries. State carefully what populations your inference compares. We would like to draw conclusions about all in-line skaters, but we have data only for injured skaters.

19.5 Exercise 19.3 describes a study that compared the proportions of athletes who use illegal drugs in two similar high schools, one that tests for drugs and one that does not. Drug testing is intended to reduce use of drugs. Select **Stat ➤ Basic Statistics ➤ 2 Proportions** to see if the data give good reason to think that drug use among athletes is lower in schools that test for drugs. (State hypotheses, and find the test statistic and the *P*-value. Be sure to state your conclusion. Because the study is not an experiment, the conclusion depends on the assumption that athletes in these two schools can be considered SRSs from all similar schools.)

19.9 Does involving a statistician to help with statistical methods improve the chance that a medical research paper will be published? A study of papers submitted to two medical journals found that 135 of 190 papers that lacked statistical assistance were rejected without even being reviewed in detail. In contrast, 293 of the 514 papers with statistical help were sent back without review.

(a) Select **Stat ➤ Basic Statistics ➤ 2 Proportions** to see if there is a significant difference in the proportions of papers with and without statistical help that are rejected without review. (This observational study does not establish causation: the studies that include statistical help may also be better than those that do not in other ways.)

(b) Select **Stat ➤ Basic Statistics ➤ 1 Proportion** to give a 95% confidence interval for the proportion of papers submitted to these journals that include help from a statistician.

(c) Select **Stat ➤ Basic Statistics ➤ 2 Proportions** to give a 95% confidence interval for the difference between the proportions of papers rejected without review when a statistician is and is not involved in the research.

19.12 A sample survey asked 202 black parents and 201 white parents of high school children "Are the public high schools in your state doing an excellent, good, fair or poor job, or don't you know enough to say?" The investigators suspected that black parents are generally less satisfied with their public schools than are whites. Among the black parents, 81 thought high schools were doing a "good" or "excellent" job; 103 of the white parents felt this way. Select **Stat ➤ Basic Statistics ➤ 2 Proportions** to see if there is good evidence that the proportion of all black parents who think their state's high schools are good or excellent is lower than the proportion of white parents with this opinion.

19.13 The sample survey described in the previous exercise also asked respondents if they agreed with the statement, "A college education has become as important as a high school diploma used to be." In the sample, 125 of 201 white parents and 154 of 202 black parents said that they "strongly agreed." Is there good reason to think that different percents of all black and white parents would strongly agree with the statement? Select **Stat ➤ Basic Statistics ➤ 2 Proportions** to find out.

19.14 The proportion of drivers who use seat belts depends on things like age (young people are more likely to go unbelted) and gender (women are more likely to use belts). It also depends on local law. In New York City, police can stop a driver who is not belted. In Boston (as of late 2000), police can cite a driver for not wearing a belt only if the driver has been stopped for some other violation. Here are data from observing random samples of female Hispanic drivers in these two cities:

City	Drivers	Belted
New York	220	183
Boston	117	68

(a) Is this an experiment or an observational study? Why?

(b) Comparing local law suggests the hypothesis that a smaller proportion of drivers wear seat belts in Boston than in New York. Select **Stat ➤ Basic Statistics ➤ 2 Proportions** to see if the data give good evidence that this is true for female Hispanic drivers.

19.15 Here are data from the study described in the previous exercise for Hispanic and white male drivers in Chicago:

Group	Drivers	Belted
Hispanic	539	286
White	292	164

Is there a significant difference between Hispanic and white drivers? How large is the difference? Select **Stat ➤ Basic Statistics ➤ 2 Proportions** to do inference to answer both questions. Be sure to explain exactly what inference you choose to do.

19.17 A study by the National Athletic Trainers Association surveyed 1679 high school freshmen and 1366 high school seniors in Illinois. Results showed that 34 of the freshmen and 24 of the seniors had used anabolic steroids. Steroids, which are dangerous, are sometimes used to improve athletic performance.

(a) To draw conclusions about all Illinois freshmen and seniors, how should the study samples be chosen?

(b) Give a 95% confidence interval for the proportion of all high school freshmen in Illinois who have used steroids.

(c) Is there a significant difference between the proportions of freshman and seniors who have used steroids?

19.20 North Carolina State University looked at the factors that affect the success of students in a required chemical engineering course. Students must get a C or better in the course in order to continue as chemical engineering majors. There were 65 students from urban or suburban backgrounds, and 52 of these students succeeded. Another 55 students were from rural or small-town backgrounds; 30 of these students succeeded in the course.

(a) Is there good evidence that the proportion of students who succeed is different for urban/suburban versus rural/small-town backgrounds? State the hypotheses. Select **Stat ➤ Basic Statistics ➤ 2 Proportions** to find the *P*-value of a test. State your conclusion.

(b) Select **Stat ➤ Basic Statistics ➤ 2 Proportions** to give a 90% confidence interval for the size of the difference.

19.25 Never forget that even small effects can be statistically significant if the samples are large. To illustrate this fact consider the study of small-business failures. Of 148 food-and-drink businesses in central Indiana 106 were headed by men and 42 were headed by women. During a three-year period 15 of the men's businesses and 7 of the women's businesses failed

(a) Find the proportions of failures for businesses headed by women and businesses headed by men. These sample proportions are quite close to each other. Select **Stat ➤ Basic Statistics ➤ 2 Proportions** to give the *P*-value for the *z* test of the hypothesis that the same proportion of women's and men's businesses fail. (Use the two-sided alternative.) The test is very far from being significant.

(b) Now suppose that the same sample proportions came from a sample 30 times as large. That is, 210 out of 1260 businesses headed by women and 450 out of 3180 businesses headed by men fail. Verify that the proportions of failures are exactly the same as in (a). Select **Stat ➤ Basic Statistics ➤ 2 Proportions** to repeat the *z* test for the new data, and show that it is now significant at the $\alpha = 0.05$ level.

(c) It is wise to use a confidence interval to estimate the size of an effect, rather than just giving a *P*-value. Select **Stat ➤ Basic Statistics ➤ 2 Proportions** to give a 95% confidence intervals for the difference between the proportions of women's and men's businesses that fail for the settings of both (a) and (b). What is the effect of larger samples on the confidence interval?

19.26 Do drivers reduce excessive speed when they encounter police radar? The EESEE story "Radar Detectors and Speeding" describes a study of the behavior of drivers on a rural interstate highway in Maryland where the speed limit was 55 miles per hour. They measured speed with an electronic device hidden in the pavement and, to eliminate large trucks, considered only vehicles less than 20 feet long. During some time periods, police radar was set up at the measurement location. Here are some of the data:

	Number of vehicles	Number over 65 mph
No radar	12,931	5,690
Radar	3,285	1,051

(a) Select **Stat ➤ Basic Statistics ➤ 2 Proportions** to give a 95% confidence interval for the proportion of vehicles going faster than 65 miles per hour when no radar is present.

(b) Select **Stat ➤ Basic Statistics ➤ 2 Proportions** to give a 95% confidence interval for the effect of radar, as measured by the difference in proportions of vehicles going faster than 65 miles per hour with and without radar.

(c) The researchers chose a rural highway so that cars would be separated rather than in clusters where some cars might slow because they see other cars slowing. Explain why such clusters might make inference invalid.

Chapter 20
Two Categorical Variables: The Chi-Square Test

Topics to be covered in this chapter:

The Chi-Square Test
The Chi-Square Test for Goodness of Fit

The Chi-Square Test

Minitab does a χ^2 test of the null hypothesis that there is "no relationship" between the column variable and the row variable in a two-way table. Example 20.1 of BPS looks at the health care system in the United States and Canada. The study looked at outcomes a year after a heart attack. One outcome was the patients' own assessment of their quality of life relative to what it had been before the heart attack. The data for the patients who survived a year are in EX20-01.MTW. To obtain tables of counts and/or percents, select

Stat ➤ Tables ➤ Cross Tabulation and Chi-Square

from the menu. In the dialog box enter the variables containing the categories that define the rows and column of the table, as shown on the following page. Click OK to obtain the following summary data.

Tabulated statistics: Quality of life, Country

```
Rows: Quality of life   Columns: Country

                            United
                    Canada  States   All

About the same          96     779   875
Much better             75     541   616
Much worse              19      65    84
Somewhat better         71     498   569
Somewhat worse          50     282   332
All                    311    2165  2476

Cell Contents:       Count
```

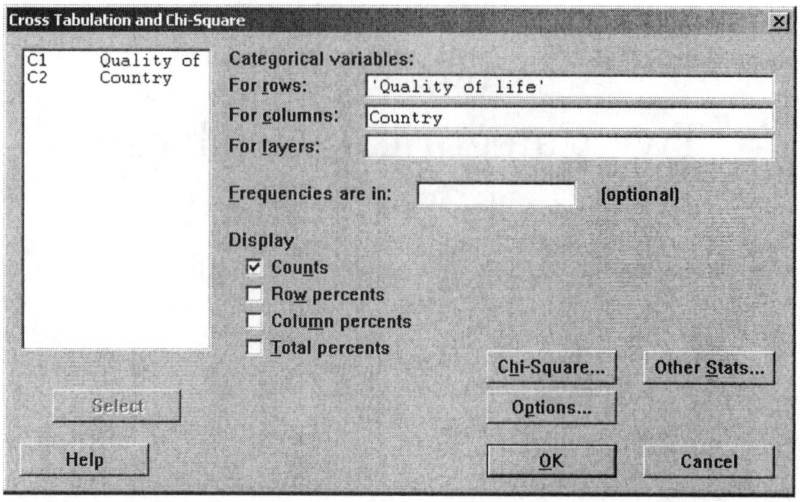

Cross Tabulation and Chi-Square

C1 Quality of
C2 Country

Categorical variables:
For rows: 'Quality of life'
For columns: Country
For layers:

Frequencies are in: (optional)

Display
☑ Counts
☐ Row percents
☐ Column percents
☐ Total percents

Chi-Square... Other Stats...
Select Options...
Help OK Cancel

To perform a chi-square test of association between variables, click on the Chi-Square button and select Chi-Square analysis in the subdialog box. You may also select 'Expected cell counts' to obtain the following output.

Tabulated statistics: Quality of life, Country

```
Rows: Quality of life   Columns: Country

                           United
                   Canada  States     All

About the same         96     779     875
                    109.9   765.1   875.0

Much better            75     541     616
                     77.4   538.6   616.0

Much worse             19      65      84
                     10.6    73.4    84.0

Somewhat better        71     498     569
                     71.5   497.5   569.0

Somewhat worse         50     282     332
                     41.7   290.3   332.0

All                   311    2165    2476
                    311.0  2165.0  2476.0

Cell Contents:        Count
                      Expected count


Pearson Chi-Square = 11.725, DF = 4, P-Value = 0.020
Likelihood Ratio Chi-Square = 10.435, DF = 4, P-Value = 0.034
```

Minitab provides the χ^2 statistic, the number of degrees of freedom, and the *P*-value. The number of degrees of freedom for the χ^2 statistic is equal to $(\text{rows}-1)\times(\text{columns}-1)$. The Minitab outputs shows that the χ^2 value is equal to 11.725, the degrees of freedom is equal to 4, and the *P*-value is equal to 0.02. There is a statistically significant relationship between patients' assessment of their quality of life and the country where they are treated for a heart attack.

Sometimes instead of raw data only summary data are available. Minitab can perform a chi-square test if the data are entered into a worksheet as shown below.

Worksheet 5 ***

↓	C1-T	C2	C3	C4	C5
	Quality of life	Canada	United States		
1	About the same	96	779		
2	Much better	75	541		
3	Much worse	19	65		
4	Somewhat better	71	498		
5	Somewhat worse	50	282		
6					
7					

To perform a Chi-square test for data in this format, select **Stat ➤ Tables ➤ Chi-Square Test (Table in Worksheet)** from the menu. Enter the columns containing the table and click OK. The columns must contain integer values. The results are the same as those on the previous page except that the rows are ordered by number instead of alphabetically as on the previous page.

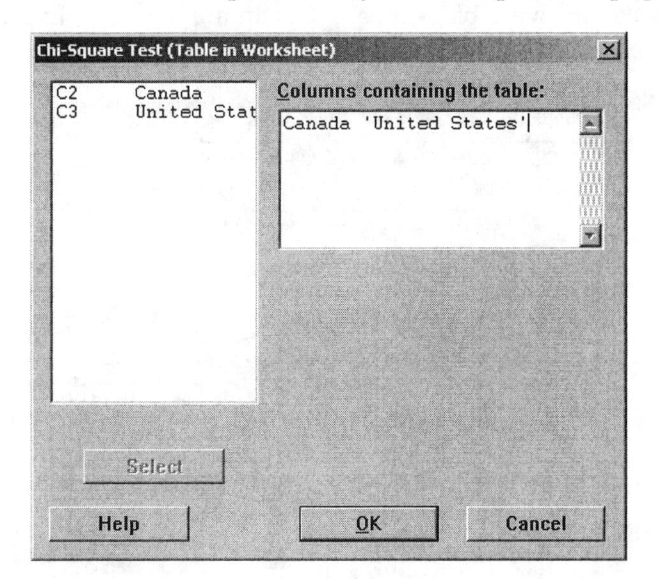

The Chi-Square Test for Goodness of Fit

The chi-square test can also be used to test that a categorical variable has a specified distribution. Example 20.9 in BPS considers the question of whether or not births are equally likely on all days of the week. The null hypothesis says that the probabilities are the same on all days. That is,

$$H_0 : p_1 = p_2 = p_3 = p_4 = p_5 = p_6 = p_7 = \frac{1}{7}$$

The alternative hypothesis is that they are not equally likely. A sample of birth records shows the following distribution.

Day	Sun.	Mon.	Tue.	Wed.	Thu.	Fri.	Sat.
Births	13	23	24	20	27	18	15

Since there were a total of 140 births, under the null hypothesis 20 births per day would be expected. Enter the observed and expected number of births into a Minitab worksheet.

Worksheet 1 ***

↓	C1	C2	C3	C4	C5	C6
	Observed	Expected				
1	13	20				
2	23	20				
3	24	20				
4	20	20				
5	27	20				
6	18	20				
7	15	20				
8						

To see if these data give significant evidence that local births are not equally likely on all days of the week, select **Calc ➤ Calculator** from the menu. In Store result in variable, type "Chisquare". In Expression, enter "SUM((Observed-Expected)**2/Expected)". Click OK to calculate the χ^2 value.

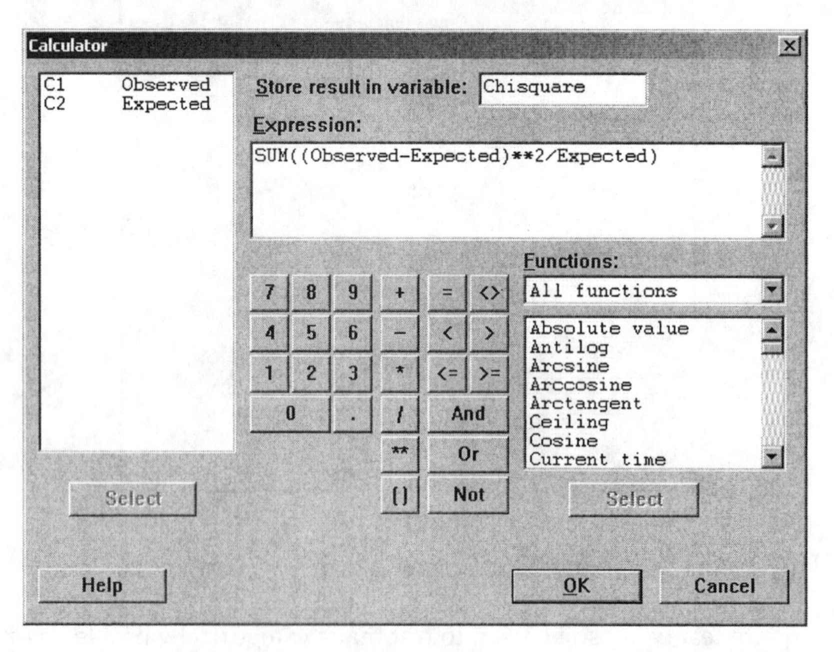

Next, choose **Calc ➤ Probability Distributions➤ Chi-Square** from the menu. Choose Cumulative probability and in Degrees of freedom type 6 (the number of outcomes minus one, 7-1=6). Enter "Chisquare" for the input column and "Cum-Prob" for the optional storage column.

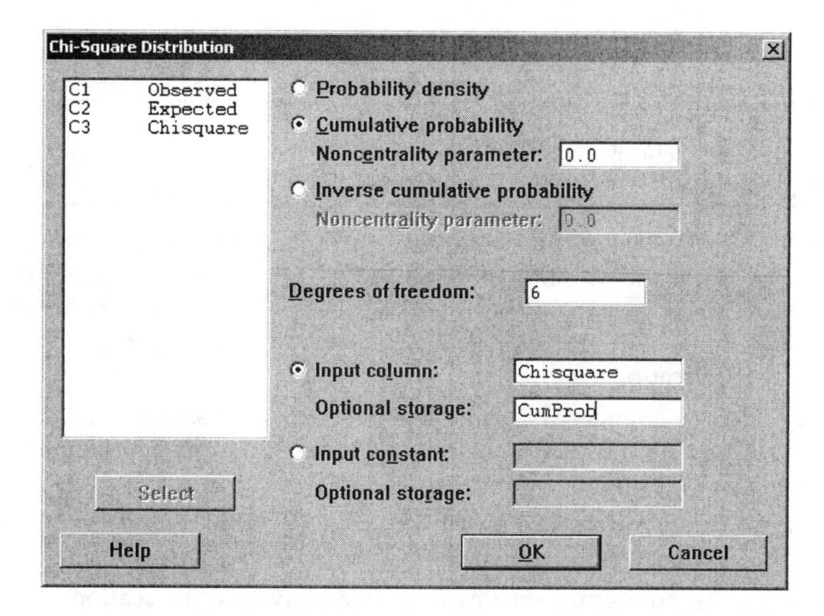

Choose **Calc ➤ Calculator** once more. In Store result in variable type "Pvalue". In Expression, enter "1-CumProb". The worksheet below shows the χ^2 statistic and *P*-value. The *P*-value is 0.269, a large value. These 140 births do not give convincing evidence that births are not equally likely on all days of the week.

Worksheet 1 ***

↓	C1	C2	C3	C4	C5	C6
	Observed	Expected	Chisquare	CumProb	Pvalue	
1	13	20	7.6	0.731103	0.268897	
2	23	20				
3	24	20				
4	20	20				
5	27	20				
6	18	20				
7	15	20				
8						

EXERCISES

20.1 Smoking is now more common in much of Europe than in the United States. In the United States, there is a strong relationship between education and smoking: well-educated people are less likely to smoke. Does a similar relationship hold in France? Here and in EX20-01.MTW are data giving the level of education and smoking status (nonsmoker, former smoker, moderate smoker, heavy smoker) of a sample of 459 French men aged 20 to 60 years. Consider them to be an SRS of men from their region of France. Select **Stat ➤ Tables ➤ Cross Tabulation and Chi-Square** from the menu to answer the following questions.

Education	Smoking Status			
	Nonsmoker	Former	Moderate	Heavy
Primary school	56	54	41	36
Secondary school	37	43	27	32
University	53	28	36	16

(a) What percent of men with a primary school education are nonsmokers? Former smokers? Moderate smokers? Heavy smokers? These percents should add to 100% (up to roundoff error). They form the conditional distribution of smoking given a primary education.

(b) In a similar way, find the conditional distributions of smoking among men with a secondary education and among men with a university education.

(c) Compare the three distributions. Is there any clear relationship between education and smoking?

(d) We conjecture that men with a university education smoke less than the null hypothesis calls for. Does comparing the observed and expected counts in this row agree with this conjecture?

20.10 A study of the career plans of young women and men sent questionnaires to all 722 members of the senior class in the College of Business Administration at the University of Illinois. One question asked which major within the business program the student had chosen. Here are the data from the students who responded:

	Female	Male
Accounting	68	56
Administration	91	40
Economics	5	6
Finance	61	59

(a) Enter the data into a worksheet and select **Stat ➤ Tables ➤ Chi-Square Test (Table in Worksheet)** to test the null hypothesis that there is no relation between the gender of students and their choice of major. Give a *P*-value and state your conclusion.

(b) Describe the differences between the distributions of majors for women and men with percents, with a graph, and in words.

(c) Which two cells have the largest terms in the sum that makes up the chi-square statistic? How do the observed and expected counts differ in these cells? (This should strengthen your conclusions in (b).)

(d) What percent of the students did not respond to the questionnaire? The nonresponse weakens conclusions drawn from these data.

20.12 Births really are not evenly distributed across the days of the week. The data in Example 20.9 failed to reject this null hypothesis because of random variation in a quite small number of births. Here are data on 700 births in the same locale:

Day	Sun.	Mon.	Tue.	Wed.	Thu.	Fri.	Sat.
Births	84	110	124	104	94	112	72

(a) The null hypothesis is that all days are equally probable. What are the probabilities specified by this null hypothesis? What are the expected counts for each day in 700 births?

(b) Enter the observed and expected counts into a Minitab worksheet. Select **Calc ➤ Calculator** to calculate the chi-square statistic for goodness of fit.

(c) What are the degrees of freedom for this statistic? Select **Calc ➤ Probability Distributions ➤ Chi-Square** from the menu to see if the 700 births give significant evidence that births are not equally probable on all days of the week.

20.14 The first digits of legitimate records such as invoices and expense account claims often follow the distribution known as Benford's law:

First digit	1	2	3	4	5	6	7	8	9
Probability	0.301	0.176	0.125	0.097	0.079	0.067	0.058	0.051	0.046

A purchasing manager is faking invoices to steer payments to a company owned by her brother. She cleverly generates random numbers to avoid obvious patterns. Here and in EX20-14.MTW are the counts of first digits in a sample of 45 invoices examined by an auditor:

First digit	1	2	3	4	5	6	7	8	9
Count	6	4	6	7	3	5	6	4	4

(a) The auditor tests the null hypothesis that the first digits follow Benford's law. Select **Calc ➤ Calculator** to find the expected counts for each of the 9 possible first digits in a sample of 45 invoices.

(b) Select **Calc ➤ Calculator** and **Calc ➤ Probability Distributions ➤ Chi-square** to find the χ^2 statistic and P-value for goodness of fit.

(c) Do the sample invoices give good evidence that first digits do not follow Benford's law in the population of all invoices filed by this manager? By examining the terms of the chi-square statistic, describe the most important deviations of the sample from the null hypothesis.

20.16 The study described in Example 20.1 in BPS also asked the patients to rate their physical capacity a year after their heart attack, relative to what it was before the attack. Here and in EX20-16.MTW are the responses:

Physical capacity	Canada	United States
Much better	37	325
Somewhat better	56	325
About the same	109	1039
Somewhat worse	78	390
Much worse	31	86
Total	311	2165

Select **Stat ➤ Tables ➤ Cross Tabulation and Chi-Square** to do both data analysis and a formal test to compare the distributions of outcomes in the two countries. Write a clear summary of your findings.

20.17 Shopping at secondhand stores is becoming more popular and has even attracted the attention of business schools. A study of customers' attitudes toward secondhand stores interviewed samples of shoppers at two secondhand stores of the same chain in two cities. Here is the two-way table comparing the income distributions in the two stores:

Income	City 1	City 2
Under $10,000	70	62
$10,000 to $19,999	52	63
$20,000 to $24,999	69	50
$25,000 to $34,999	22	19
$35,000 or more	28	24

Enter the data into a worksheet and select **Stat ➤ Tables ➤ Chi-Square Test** to calculate the chi-square statistic χ^2, the degrees of freedom, and the *P*-value. Is there good evidence that customers at the two stores have different income distributions?

20.18 A large study of child care used samples from the data tapes of the Current Population Survey over a period of several years. The result is close to an SRS of child-care workers. The Current Population Survey has three classes of child-care workers: private household, nonhousehold, and preschool teacher. Here are data on the number of blacks among women workers in these three classes:

	Total	Black
Household	2455	172
Nonhousehold	1191	167
Teachers	659	86

(a) Enter the data into a worksheet. Select **Calc ➤ Calculator** to find the percent of each class of child-care workers that is black.

(b) Use the calculator to make a two-way table of class of worker by race (black or other).

(c) Can we safely use the chi-square test? What null and alternative hypotheses does χ^2 test?

(d) Select **Stat ➤ Tables ➤ Chi-Square Test (Table in Worksheet)** to calculate the chi-square statistic χ^2, the degrees of freedom, and the *P*-value for this table.

(e) What do you conclude from these data?

20.19 Exercise 6.20 and EX06-20.MTW give these data on the rank and gender of professors at a large university:

	Female	Male	Total
Assistant professors	126	213	339
Associate professors	149	411	560
Professors	60	662	722
Total	335	1286	1621

Women are underrepresented at the full professor level. Enter the data into a Minitab worksheet. Select **Stat ➤ Tables ➤ Chi-Square Test (Table in Worksheet)** to see if there is a significant difference between the distribution of ranks for female and male faculty? Which cells contribute the most to the overall chi-square statistic?

20.20 "Do you think there should be a law that would ban possession of handguns except for the police and other authorized persons?" Exercise I.14 of BPS and EX20-20.MTW give these data on the responses of a random sample of adults, broken down by level of education:

Education	Yes	No
Less than high school	58	58
High school graduate	84	129
Some college	169	294
College graduate	98	135
Postgraduate degree	77	99

Select **Stat ➤ Tables ➤ Chi-Square (Table in Worksheet)** from the menu to carry out a chi-square test, giving the statistic χ^2 and its *P*-value. How strong is the evidence that people with different levels of education feel differently about banning private possession of handguns?

20.21 How are the smoking habits of students related to their parents' smoking? Here and in EX06-16.MTW are data from eight high schools on smoking among students and among their parents:

	Neither parent smokes	One parent smokes	Both parents smoke
Student does not smoke	1168	1823	1380
Student smokes	188	416	400

Select **Stat ➤ Tables ➤ Chi-Square (Table in Worksheet)** from the menu to answer the following questions.

(a) Find the percent of students who smoke in each of the three parent groups. Make a graph to compare these percents. Describe the association between parent smoking and student smoking.

(b) Explain in words what the null hypothesis for the chi-square test says about student smoking.

(c) Find the expected counts if H_0 is true, and display them in a two-way table similar to the table of observed counts.

(d) Compare the tables of observed and expected counts. Explain how the comparison expresses the same association you saw in (a).

(e) Give the chi-square statistic and its *P*-value. Examine the terms of chi-square to confirm the pattern you saw in (a) and (d). What is your overall conclusion?

20.25 Exercise I.13 of BPS and EX20-25.MTW give these data on the responses of random samples of black, Hispanic and white parents to the question "Are the high schools in your state doing an excellent, good, fair or poor job, or don't you know enough to say?"

	Black parents	Hispanic parents	White parents
Excellent	12	34	22
Good	69	55	81
Fair	75	61	60
Poor	24	24	24
Don't know	22	28	14
Total	202	202	201

Select **Stat ➤ Tables ➤ Chi-Square (Table in Worksheet)** from the menu to see if the differences in the distributions of responses for the three groups of parents are statistically significant. What departures from the null hypothesis "no relationship between group and response" contribute most to the value of the chi-square statistic? Write a brief conclusion based on your analysis.

20.28 Cancer of the colon and rectum is less common in the Mediterranean region than in other Western countries. The Mediterranean diet contains little animal fat and lots of olive oil. Italian researchers compared 1953 patients with colon or rectal cancer with a control group of 4154 patients admitted to the same hospitals for unrelated reasons. They estimated consumption of various foods from a detailed interview, then divided the patients into three groups according to their consumption of olive oil. Here and in EX20-28.MTW are some of the data:

	Olive Oil			
	Low	Medium	High	Total
Colon cancer	398	397	430	1225
Rectal cancer	250	241	237	728
Controls	1368	1377	1409	4154

(a) Is this study an experiment? Explain your answer.

(b) Is high olive oil consumption more common among patients without cancer than in patients with colon cancer or rectal cancer?

(c) Select **Stat ➤ Tables ➤ Chi-Square (Table in Worksheet)** from the menu to find the chi-square statistic χ^2. What would be the mean of χ^2 if the null hypothesis (no relationship) were true? What does comparing the observed value of χ^2 with this mean suggest? What is the P-value? What do you conclude?

(d) The investigators report that "less than 4% of cases or controls refused to participate." Why does this fact strengthen our confidence in the results?

20.31 In 1912 the luxury liner *Titanic,* on its first voyage across the Atlantic, struck an iceberg and sank. Some passengers got off the ship in lifeboats, but many died. Think of the *Titanic* disaster as an experiment in how the people of that time behaved when faced with death in a situation where only some can escape. The passengers are a sample from the population of their peers. Here and in EX20-31.MTW are data about who lived and who died, by gender and economic status. Select **Stat ➤ Tables ➤ Cross Tabulation and Chi-Square** to answer the following questions.

Men

Status	Died	Survived
Highest	111	61
Middle	150	22
Lowest	419	85
Total	680	168

Women

Status	Died	Survived
Highest	6	126
Middle	13	90
Lowest	107	101
Total	126	317

(a) Compare the percents of men and of women who died. Is there strong evidence that a higher proportion of men die in such situations? Why do you think this happened?

(b) Select **Data ➤ Unstack Columns** to unstack the data by gender. Look only at the women. Describe how the three economic classes differ in the percent of women who died. Are these differences statistically significant?

(c) Now look only at the men and answer the same questions.

20.33 Exercise 20.17 describes a study of shoppers at secondhand stores in two cities. The breakdown of the respondents by gender is as follows:

	City 1	City 2
Men	38	68
Women	203	150
Total	241	218

Is there a significant difference between the proportions of women customers in the two cities?

(a) State the null hypothesis. Select **Stat ➤ Basic Statistics ➤ 2 Proportions** to do a two-sided z test, and give a P-value.

(b) Enter the data into a worksheet and select **Stat ➤ Tables ➤ Chi-Square Test (Table in Worksheet)** to calculate the chi-square statistic χ^2 and show that it is the square of the z statistic. Show that the P-value agrees with your result from (a).

(c) Select **Stat ➤ Basic Statistics ➤ 2 Proportions** to give a 95% confidence interval for the difference between the proportions of women customers in the two cities.

Chapter 21
Inference for Regression

Topics to be covered in this chapter:

Estimating the Regression Parameters
Confidence Intervals and Hypothesis Tests for β
Inference for Correlation
Inference about Prediction

Estimating the Regression Parameters

In Example 21.1 of BPS examines the relationship between the crying of infants four to ten days old and their later IQ test scores. A snap of a rubber band on the sole of the foot caused the infants to cry. The researchers recorded the crying and measured its intensity by the number of peaks in the most active 20 seconds. They later measured the children's IQ at age three years using the Stanford-Binet IQ test. Table 21.1 and TA21-01.MTW contain data on 38 infants. We begin with data analysis, following the usual steps.

Before attempting inference, examine the data by (1) making a scatterplot, (2) fitting the least-squares regression, $\hat{y} = b_0 + b_1 x$, (3) checking for outliers and influential observations, and (4) computing the value of r^2. These can all be done at once by making a fitted line plot. Fitted line plots can be obtained by selecting

Stat ➤ Regression ➤ Fitted line plot

from the menu and entering the appropriate Predictor and Response variable.

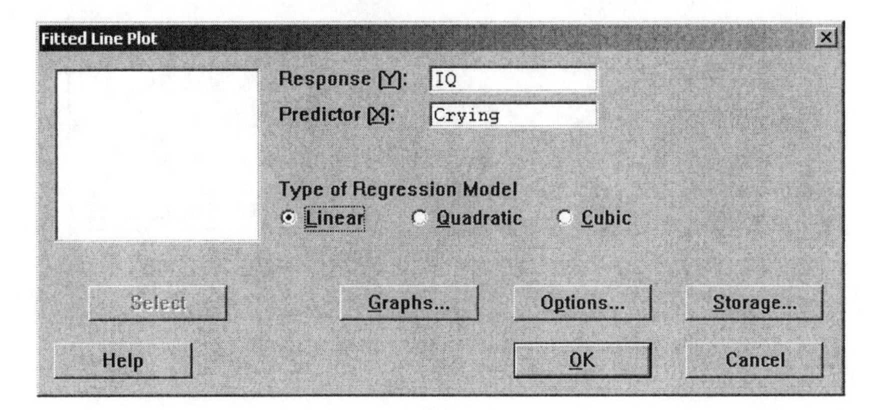

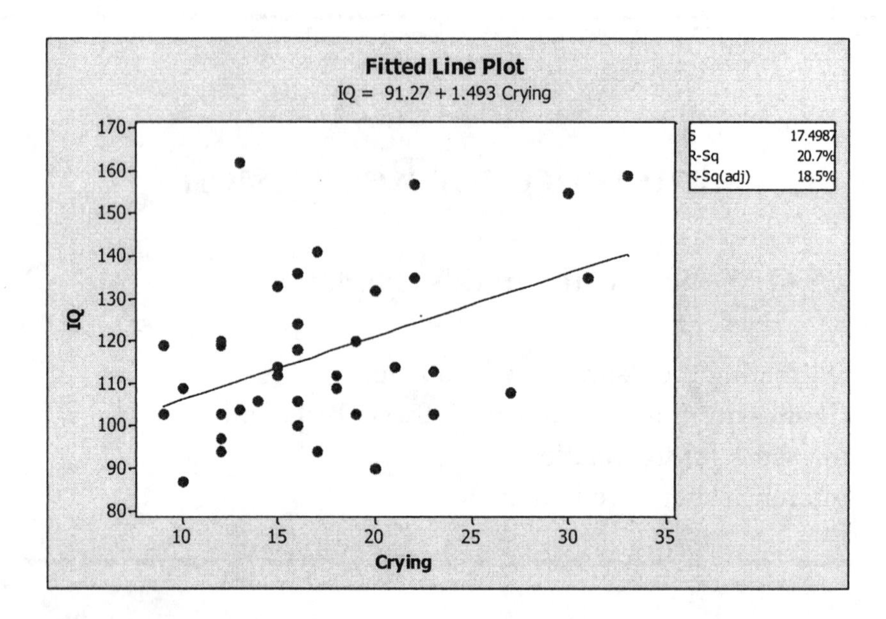

The scatterplot shows a moderate linear relationship with no extreme outliers or potentially influential observations. The least-squares line is given to be

$$\text{IQ} = 91.27 + 1.493 \text{ Crying}$$

and $r^2 = 0.207$. Because $r^2 = 0.207$, only about 21% of the variation in IQ scores is explained by crying intensity.

To use the regression line for prediction, select

Stat ➤ Regression ➤ Regression

from the menu. The Response variable (IQ) and the Predictor variable (Crying) are entered in the dialog box.

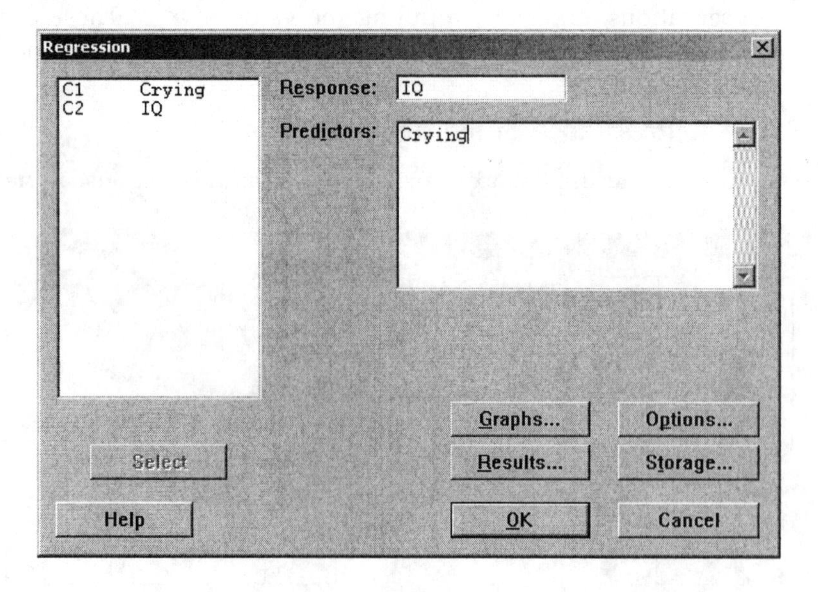

Regression Analysis: IQ versus Crying

```
The regression equation is
IQ = 91.3 + 1.49 Crying

Predictor     Coef  SE Coef      T      P
Constant    91.268    8.934  10.22  0.000
Crying      1.4929   0.4870   3.07  0.004


S = 17.4987   R-Sq = 20.7%   R-Sq(adj) = 18.5%


Analysis of Variance

Source          DF       SS      MS     F      P
Regression       1   2877.5  2877.5  9.40  0.004
Residual Error  36  11023.4   306.2
Total           37  13900.9


Unusual Observations

Obs  Crying      IQ     Fit  SE Fit  Residual  St Resid
 32    31.0  135.00  137.55    7.21     -2.55    -0.16 X
 37    33.0  159.00  140.53    8.11     18.47     1.19 X
 38    13.0  162.00  110.68    3.56     51.32     3.00R

R denotes an observation with a large standardized residual.
X denotes an observation whose X value gives it large influence.
```

The values of a and b are given in the column labeled Coef. The column labeled Predictor tells us that the first entry (Constant) is a, the intercept, and the second is b, the slope. We see that $a = 91.268$ and $b = 1.4929$. These are the estimates of α and β. These values are rounded and appear in the regression equation

$$\text{IQ} = 91.27 + 1.493\ \text{Crying}$$

The standard error, $s = 17.4987$, is used to estimate σ, the standard deviation of responses about the true regression line.

In this example, we have fitted a line and we should now examine the residuals. To obtain residual plots, click on the Graphs button in the Regression dialog box. In the subdialog box, enter "Crying" to obtain a plot of the residuals versus the explanatory variable. You can also place a check next to Histogram of the residuals to check for normality.

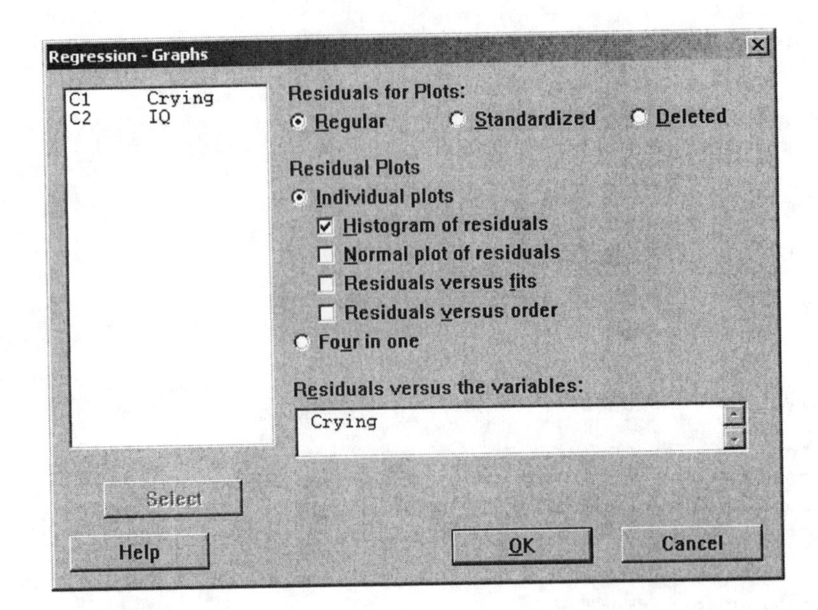

The residual plot for these data looks satisfactory.

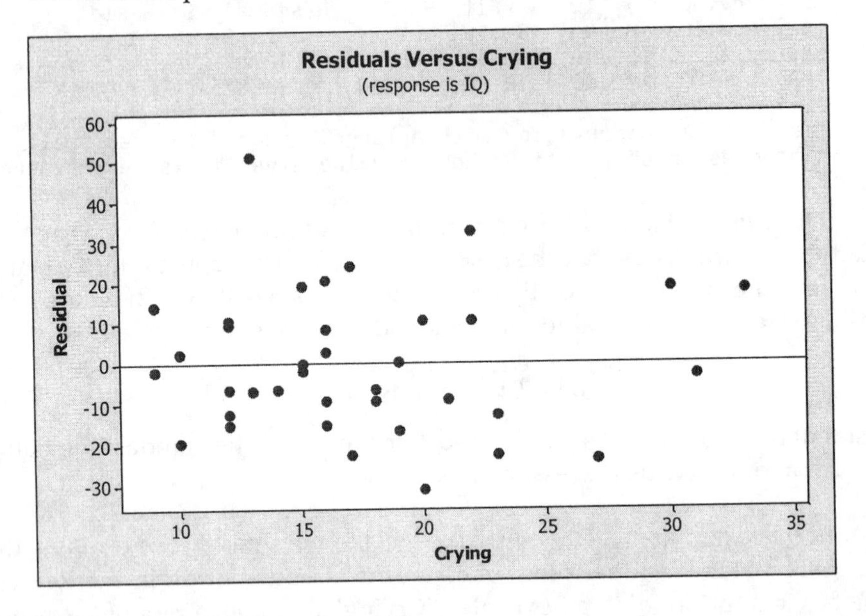

The histogram of the residuals shows an outlier, but it does not appear to be a serious departure from Normality. This is important for the inference that follows.

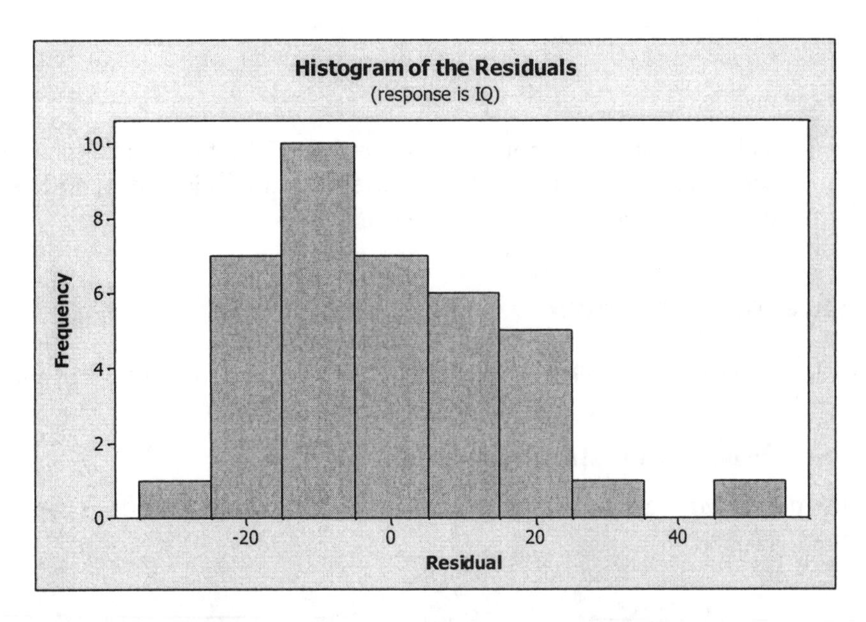

Confidence Intervals and Hypothesis Tests for β

Confidence intervals and tests for the slope and intercept are based on the normal sampling distributions of the estimates b_0 and b_1. Since the standard deviations are not known, a t distribution is used. The value of SE_b is 0.4870. It appears in the output from the regression to the right of the estimated slope $b = 14929$. Similarly, the value of SE_a is 8.934. It appears to the right of the estimated constant, $a = 91.264$. Confidence intervals for α and β have the form

$$\text{estimate} \pm t^* \text{SE}_{\text{estimate}}$$

The value of t^* can be calculated using **Calc ➤ Probability Distributions ➤ *t*.** The t distributions for this problem have $n - 2 = 36$ degrees of freedom. In the dialog box, select Inverse cumulative probability and enter 36 degrees of freedom. For a 95% confidence interval enter 0.025 as the Input constant.

Inverse Cumulative Distribution Function

```
Student's t distribution with 36 DF

P( X <= x )          x
      0.025   -2.02809
```

Therefore, the value of t^* is 2.028. The upper and lower bounds for the confidence interval can be calculated with Minitab's calculator. A 95% confidence interval for β is (0.505220, 2.48058).

The t statistic and P-value for the test of

$$H_0: \beta = 0$$
$$H_a: \beta \neq 0$$

appear in the columns labeled T and P. The t ratio can also be obtained from the formula

$$t = \frac{b}{SE_b} = \frac{1.4929}{0.4870} = 3.0655$$

The P-value is listed as 0.004. There is strong evidence that there is a relationship between IQ scores and crying intensity. Confidence intervals and hypothesis tests for α can be obtained similarly.

Inference for Correlation

To calculate the correlation coefficient and do a two-tailed test of the correlation, select

Stat ➤ Basic Statistics ➤ Correlation

from the menu. Minitab calculates r, the sample correlation, and does the significance test

$$H_0: \rho = 0$$
$$H_1: \rho \neq 0$$

The variables can be entered in the dialog box in any order. Check Display p-values for the test and click OK.

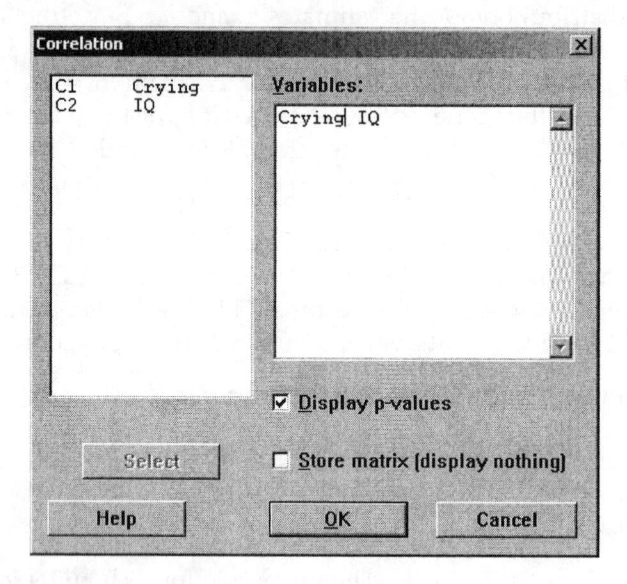

Notice that the P-value that follows is exactly the same as the P-value obtained for the slope. This is always the case.

Correlations: Crying, IQ

```
Pearson correlation of Crying and IQ = 0.455
P-Value = 0.004
```

Inference about Prediction

We found that the least-squares line for predicting IQ from 'Crying' is

$$\hat{y} = 91.27 + 1.493x.$$

Minitab can be used to predict the IQ for an infant who had 10 crying peaks. Select **Stat ➤ Regression ➤ Regression** and click on the Options button. Specify a predicted value in the Options subdialog box.

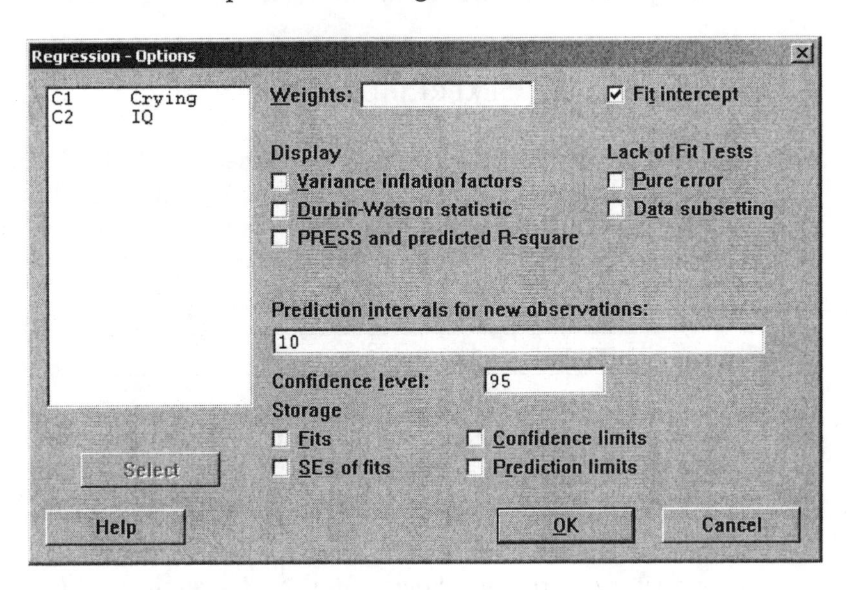


```
New
Obs      Fit   SE Fit         95% CI              95% PI
  1   106.20     4.59   (96.90, 115.50)   (69.51, 142.88)


Values of Predictors for New Observations

New
Obs   Crying
  1     10.0
```

We may be interested in predicting the *mean response,* the average IQ for all infants with 10 crying peaks, or we may be interested in predicting the IQ of *one individual* who had 10 crying peaks. The prediction is the same for both, $\hat{y} = 106.2$. However, the margin of error is different for the two kinds of prediction.

Individual infants with 10 crying peaks don't all have the same IQ. So we need a larger margin of error to pin down one infant's IQ than to estimate the mean IQ for all infants who had 10 crying peaks. The interval for the *mean* is listed under 95% CI as (96.90, 115.50). The interval is $\hat{y} \pm t^* \text{SE}_{\hat{\mu}}$ and the value labeled SE Fit (4.59) is $\text{SE}_{\hat{\mu}}$. The *individual* prediction interval is listed under 95% PI as (69.51, 142.88). This interval is $\hat{y} \pm t^* \text{SE}_{\hat{y}}$. The value of $\text{SE}_{\hat{y}}$ is not given on the Minitab output, but it is easily obtained from the following formula

$$SE_{\hat{y}} = \sqrt{s^2 + (SE_{\hat{\mu}})^2}$$

Remember that before using regression inference, the data must satisfy the regression model assumptions. Use a scatterplot to check that the true relationship is linear. The scatter of the data points about the line should be roughly the same over the entire range of the data. A plot of the residuals against x should not show any pattern. A histogram or stemplot of the residuals should not show any major departures from normality.

EXERCISES

21.1 Archaeopteryx is an extinct beast having feathers like a bird and teeth and a long bony tail like a reptile. Here are the lengths in centimeters of the femur (a thigh bone) and the humerus (a bone in the upper arm) for the five fossil specimens that preserve both bones:

Femur	38	56	59	64	74
Humerus	41	63	70	72	84

The strong linear relationship between the lengths of the two bones helped persuade scientists that all five specimens belonged to the same species.

(a) Enter the data into a Minitab worksheet. Select **Graph ➤ Scatterplot** from the menu to make a scatterplot with femur length as the explanatory variable. Select **Stat ➤ Regression ➤ Regression** from the menu to find the equation of the least-squares regression line. Do you think that femur length will allow good prediction of humerus length?

(b) Explain in words what the slope β of the true regression line says about *Archaeopteryx*. What is the estimate of β from the data? What is your estimate of the intercept α of the true regression line? What is your estimate of the standard deviation σ in the regression model? You have now estimated all three parameters in the model.

21.2 The rate at which an icicle grows depends on temperature, water flow, and wind. The data in EX21-02.MTW and here are for an icicle grown in a cold chamber at −11°C with no wind and a water flow of 11.9 milligrams per second.

Time (min)	10	20	30	40	50	60	70	80	90
Length (cm)	0.6	1.8	2.9	4.0	5.0	6.1	7.9	10.1	10.9
Time (min)	100	110	120	130	140	150	160	170	180
Length (cm)	12.7	14.4	16.6	18.1	19.9	21.0	23.4	24.7	27.8

(a) Select **Stat ➤ Regression ➤ Fitted Line Plot** to plot the regression line and scatterplot suitable for predicting length from time. The pattern is very linear. What is the squared correlation r^2? Time explains almost all of the change in length.

(b) The model for regression inference has three parameters, which we call α, β, and σ. From the output, what are the estimates of these parameters?

(c) What is the t statistic for testing H_0: $\beta = 0$?

(d) How many degrees of freedom does t have? What is the P-value of t against the *one-sided* alternative H_a: $\beta > 0$. What do you conclude?

21.3 One effect of global warming is to increase the flow of water into the Arctic Ocean from rivers. Such an increase might have major effects on the world's climate. Six rivers (Yenisey, Lena, Ob, Pechora, Kolyma, and Severnaya Dvina) drain two-thirds of the Arctic in Europe and Asia. Several of these are among the largest rivers on earth. Table 21.2 and TA21-02.MTW contain the total discharge from these rivers each year from 1936 to 1999. Discharge is measured in cubic kilometers of water.

(a) Select **Stat ➤ Regression ➤ Fitted Line Plot** to make a scatterplot with the regression line of river discharge against time. Is there a clear increasing trend? Find r^2 and briefly interpret its value.

(b) Is the increasing trend visible in your plot statistically significant? If so, changes in the Arctic may already be affecting earth's climate. Select **Stat ➤ Regression ➤ Regression** to answer this question. Give a test statistic, its P-value, and the conclusion you draw from the test.

21.4 Exercise 21.2 and EX21-02.MTW give data on the growth of an icicle. Select **Stat ➤ Regression ➤ Regression** from the menu to find the regression output. We want a 95% confidence interval for the slope of the true regression line. Starting from the information in the Minitab output, find this interval. Say in words what the slope of the true regression line tells us about the growth of icicles under the conditions of this experiment.

21.6 Does how long young children remain at the lunch table help predict how much they eat? Here and in EX 21-06.MTW are data on 20 toddlers observed over several months at a nursery school. "Time" is the average number of minutes a child spent at the table when lunch was served. "Calories" is the average number of calories the child consumed during lunch, calculated from careful observation of what the child ate each day.

Time	21.4	30.8	37.7	33.5	32.8	39.5	22.8	34.1	33.9	43.8
Calories	472	498	465	456	423	437	508	431	479	454
Time	42.4	43.1	29.2	31.3	28.6	32.9	30.6	35.1	33.0	43.7
Calories	450	410	504	437	489	436	480	439	444	408

Select **Graph ➤ Scatterplot** to make a scatterplot of the data. Select **Stat ➤ Regression ➤ Regression** from the menu to find the equation of the least-squares line for predicting calories consumed from time at the table. Describe briefly what the data show about the behavior of children. Then give a 95% confidence interval for the slope of the true regression line.

21.9 Exercise 4.6 of BPS and EX21-09.MTW give data on the fuel consumption of a small car at various speeds from 10 to 150 kilometers per hour. Is there evidence of straight-line dependence between speed and fuel use? Select **Stat ➤ Regression ➤ Regression** to find out. Select **Graph ➤ Scatterplot** to make a scatterplot and use it to explain the result of your test.

21.10 There is some evidence that drinking moderate amounts of wine helps prevent heart attacks. Table 4.3 of BPS and EX21-10.MTW give data on yearly wine consumption (liters of alcohol from drinking wine, per person) and yearly deaths from heart disease (deaths per 100,000 people) in 19 developed nations. Select **Stat ➤ Regression ➤ Regression** to see if there is statistically significant evidence that the correlation between wine consumption and heart disease deaths is negative.

21.11 Exercise 21.6 and EX21-06.MTW give data on the time that 20 young children remained at the lunch table and the number of calories they consumed. We might think that children eat what they need either quickly or slowly, so that there is no correlation between time at the table and calories. Select **Stat ➤ Basic Statistics ➤ Correlation** to find the correlation r for these children and the P-value to test the hypothesis of population correlation 0 against the two-sided alternative that the correlation is not 0.

21.12 Exercise 21.2 and EX21-02.MTW give data on the growth of an icicle. Analysis of the data in Exercise 21.2 shows that growth of icicles is very linear. We might want to predict the mean length of icicles after 200 minutes under the same conditions of temperature, wind, and water flow. Select **Stat ➤ Regression ➤ Regression** from the menu and click on the Options button to do the prediction.

(a) What is the predicted value for $x = 200$? What is the 95% interval we want?

(b) Change the confidence level to obtain a 99% confidence interval for the mean length of icicles after 200 minutes.

21.14 Rachel is another child at the nursery school of Exercise 21.6. Over several months, Rachel averages 40 minutes at the lunch table. Give a 95% interval to predict Rachel's average calorie consumption at lunch.

21.16 Exercise 21.2 and EX21-02.MTW give data on the growth of an icicle. Select **Stat ➤ Regression ➤ Regression** from the menu and click on the Graphs button. Make a histogram of the residuals as well as a plot of the residuals versus Time. Examine the regression model conditions one by one. This example illustrates mild violations of the regression conditions that did not prevent the researchers from doing inference.

(a) Independent observations. The data come from the growth of a single icicle, not from a different icicle at each time. Explain why this would violate the independence condition if we had data on the growth of a child rather than of an icicle. (The researchers decided that all icicles, unlike all children, grow at the same rate if the conditions are held fixed. So one icicle can stand in for a separate icicle at each time.)

(b) Linear relationship. Your plot and r^2 from Exercise 21.2 show that the relationship is very linear. Residual plots magnify effects. Plot the residuals against time. What kind of deviation from a straight line is now visible? (The deviation is clear in the residual plot, but it is very small in the original scale.)

(c) Spread about the line stays the same. Your plot in (b) shows that it does not. (Once again, the plot greatly magnifies small deviations.)

(d) Normal variation about the line. Consider the histogram of the residuals. With only 18 observations, no clear shape emerges. Does strong skewness or outliers suggest lack of Normality?

21.20 Exercise I.39 in BPS and EX21-20.MTW give data for predicting the "gate velocity" of molten metal from the thickness of the aluminum piston being cast. The data come from observing skilled workers and will be used to guide less experienced workers. Select **Stat ➤ Regression ➤ Fitted Line Plot** to make a scatterplot with the regression line drawn on your plot.

(a) Give the value of r^2 and the equation of the least-squares line.

(b) Select **Stat ➤ Regression ➤ Regression** from the menu to test the hypothesis that there is no straight-line relationship between thickness and gate velocity. Give a test statistic, *P*-value, and your conclusion. Use the output to give the 90% interval for the slope of the true regression line of gate velocity on piston thickness

(c) This time, click on the Options button to make a prediction for x = 0.5 inch. Use the output to give 90% intervals for the average gate velocity for a type of piston with thickness 0.5 inch.

21.26 Exercise 4.4 in BPS and EX21-26.MTW give data on the percent of adult sparrowhawks in a colony that return from the previous year and the number of new adults that join the colony. Select **Stat ➤ Regression ➤ Regression** and click on the Options button to predict the average number of new birds in colonies to which 60% of the birds return.

(a) Write the equation of the least-squares line. What is the predicted response for x = 60%.

(b) Which 95% interval in the output gives us a margin of error for predicting the average number of new birds?

21.27 The regression of number of new birds that join a sparrowhawk colony on the percent of adult birds in the colony that return from the previous year is an example of data that satisfy the regression model well. Select **Stat ➤ Regression ➤ Regression** from the menu and click on the Graphs button to make a histogram of the residuals and a plot of the residuals versus the percent of birds that return. Examine the regression model conditions one by one.

(a) Independent observations. Why are the 13 observations independent?

(b) Linear relationship. Select **Graph ➤ Scatterplot** to make a scatterplot of the data. Is the pattern reasonably linear? A plot of the residuals against the explanatory variable x magnifies the deviations from the least-squares line. Does the plot show any systematic pattern?

(c) Spread about the line stays the same. Does your residual plot show any systematic change in spread as x changes?

(d) Normal variation about the line. With only 13 observations, no clear shape emerges on the histogram of the residuals. Does strong skewness or outliers suggest lack of Normality?

21.29 Exercise 4.21 and EX21-29.MTW give the percent returns for the Vanguard International Growth Fund, a mutual fund that buys foreign stocks, for each year between 1982 and 2001. The data also include the returns on the Morgan Stanley EAFE index, an average of all foreign stocks against which the mutual fund measures its performance. Use software to analyze the relationship between the two sets of returns.

(a) Select **Stat ➤ Regression ➤ Fitted line plot** to make a scatterplot suitable for predicting the fund's return from the EAFE return. Describe the form, direction, and strength of the relationship and give the equation of the least-squares line. Are there any extreme outliers or observations that may be very influential?

(b) Select **Stat ➤ Regression ➤ Regression** to obtain the regression output. Give a 95% confidence interval for the slope of the true regression line. Explain in plain language what this interval tells us about how the fund performs.

(c) In 2002, the EAFE index lost 15.94%. What does your regression predict to be the 2002 return on the Vanguard International Growth Fund? Select **Stat ➤ Regression ➤ Regression** and click on the Options button to give a 95% prediction interval for the 2002 return. The actual 2002 return was −17.79%. Did the prediction interval cover this return?

(d) Select **Stat ➤ Regression ➤ Regression** again. This time click on the Graphs button to plot the residuals against the predicted. Also make a histogram of the residuals. Are there signs of any major violations of the conditions for regression inference?

21.33 Table 1.2 in BPS and EX21-33.MTW give the city and highway gas mileages for 22 models of two-seater cars. The Honda Insight, a gas-electric hybrid car, is an outlier in both the x and y directions. Exercise 5.9 (page 118) asks you to investigate the influence of the Insight on the least-squares line. The influence is not large because the Insight does not lie far from the least-squares line calculated from the other 21 cars. Now you will investigate the influence of the Insight on inference. Select **Stat ➤ Regression ➤ Regression** to carry out regression both with and without the Insight.

(a) How does the Insight influence the value of the t statistic for the regression slope and its P-value? Is the influence practically important?

(b) How does the Insight influence the 95% confidence interval for the regression slope? Is the influence practically important?

21.35 We can use the data in Table 21.4 and TA21-04.MTW to study the prediction of the weight of a perch from its length.

(a) Select **Graph ➤ Scatterplot** to make a scatterplot of weight versus length, with length as the explanatory variable. Describe the pattern of the data and any clear outliers.

(b) It is more reasonable to expect the one-third power of the weight to have a straight-line relationship with length than to expect weight itself to have a straight-line relationship with length. Explain why this is true. (Hint: What happens to weight if length, width, and height all double?)

(c) Select **Calc ➤ Calculator** to create a new variable that is the one-third power of weight. Store the result in variable "weight3" and enter the expression "weight**(1/3)". Make a scatterplot of this new response variable against length. Describe the pattern and any clear outliers.

(d) Is the straight-line pattern in (c) stronger or weaker than that in (a)? Select **Stat ➤ Regression ➤ Fitted Line Plot** to compare the plots and also the values of r^2.

(e) Select **Stat ➤ Regression ➤ Regression** from the menu to find the least-squares regression line to predict the new weight variable from length. Click on the Options button to predict the mean of the new variable for perch 27 centimeters long, and give a 95% confidence interval.

(f) Examine the residuals from your regressions. Does it appear that any of the regression assumptions are not met?

Chapter 22
One-Way Analysis of Variance: Comparing Several Means

Topic to be covered in this chapter:

One-Way Analysis of Variance

One-Way Analysis of Variance

Minitab can perform a one-way analysis of variance test to compare means of different populations. The response variable must be numeric. The data can be entered with each population in separate columns on a worksheet (*unstacked* case), or with the response data in one column and another column identifying the population (*stacked* case). To perform a one-way analysis of variance with stacked data, choose

Stat ➤ ANOVA ➤ One-way

from the Minitab menu. To perform a one-way analysis of variance with unstacked data, choose

Stat ➤ ANOVA ➤ One-way (unstacked)

from the menu.

Example 22.1 in BPS considers data on the highway gas mileage (in miles per gallon) for 31 midsize cars, 31 SUVs, and 14 standard size pickup trucks. The gas mileages and vehicle classifications are compiled by the Environmental Protection Agency. The data are given in TA22-01.MTW. The data are arranged with one row for each vehicle. The vehicle type (1 = car, 2 = truck, and 3 = SUV) is given in one column and the vehicle's mileage is given in a second column. To analyze these data, select **Stat ➤ ANOVA ➤ One-way** from the menu. In the dialog box enter the column containing the MPG in the box for Response and the column containing the vehicle type in the box for Factor, then click OK.

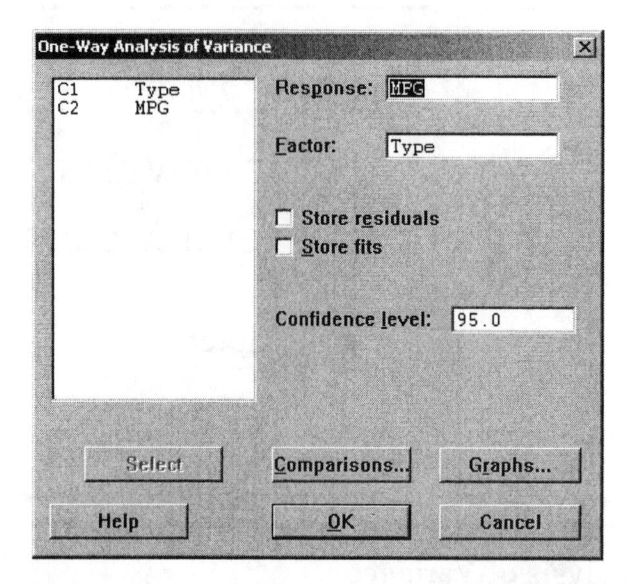

It is important to check that the assumptions of one-way analysis of variance are satisfied. Specifically, the populations are normal with possibly different means and the same variance. Boxplots are useful for visually checking these assumptions. To produce side-by-side boxplots for this data, click on the Graphs button on the One-way analysis of variance dialog box. In the Graphs subdialog box, check Boxplots of the data.

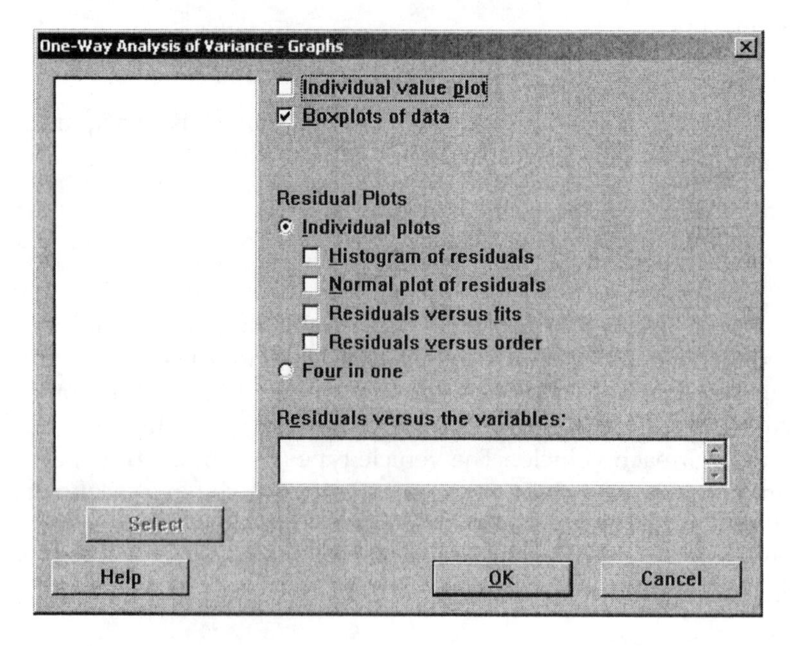

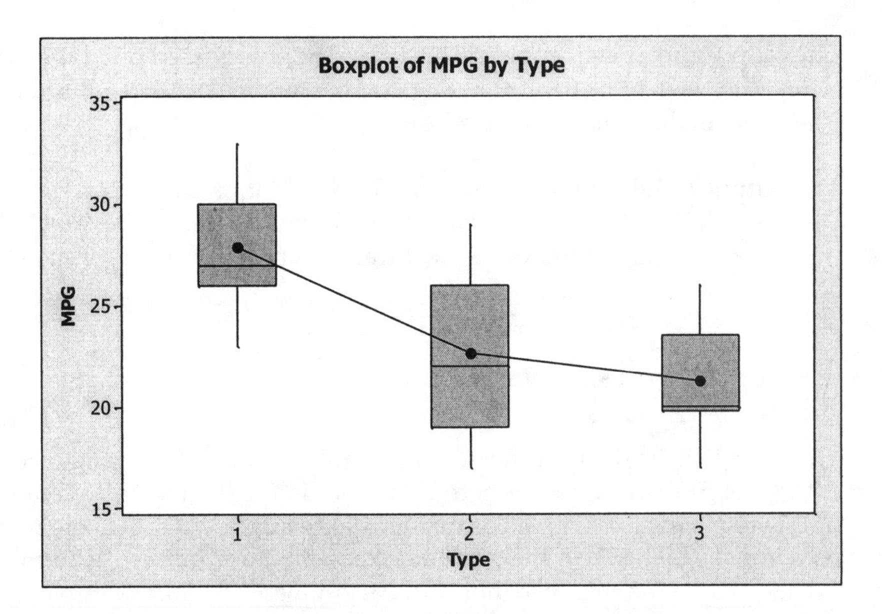

Call the mean test scores for the three educational approaches μ_1, μ_2, and μ_3. We want to test the null hypothesis that there are *no differences* among the test scores for the three groups:

$$H_0\text{: } \mu_1 = \mu_2 = \mu_3$$

The alternative is that there is some difference.

$$H_a\text{: not all of } \mu_1, \mu_2, \text{and } \mu_3 \text{ are equal}$$

The analysis of variance (ANOVA) output follows.

One-way ANOVA: MPG versus Type

```
Source  DF       SS      MS      F      P
Type     2   606.45  303.22  31.61  0.000
Error   73   700.34    9.59
Total   75  1306.79

S = 3.097   R-Sq = 46.41%   R-Sq(adj) = 44.94%


                                Individual 95% CIs For Mean Based on
                                Pooled StDev
Level   N    Mean  StDev   -+---------+---------+---------+--------
1      31  27.903  2.561                                (----*---)
2      31  22.677  3.673            (----*---)
3      14  21.286  2.758   (-----*------)
                           -+---------+---------+---------+--------
                           20.0      22.5      25.0      27.5

Pooled StDev = 3.097
```

Since the ratio of the largest to the smallest standard deviation is less than 2, it is safe to do one-way analysis of variance.

The output provides the ANOVA table. The columns in this table are labeled Source, DF (degrees of freedom), SS (sum of squares), MS (mean square), F, and P. The rows in the table are labeled Method, Error, and Total.

The output provides the pooled standard deviation in the last line. It is given as equal to 3.097. Note that it can also be computed from the ANOVA table using the sum of squares and degrees of freedom for the Error row. That is,

$$s_p^2 = \frac{\text{SS}}{\text{DF}} = \frac{700.34}{73} = 9.59$$

which implies that $s_p = 3.097$.

The F statistic is given in the ANOVA table. If H_0 is true, the F statistic has an F(DFG,DFE) distribution, where DFG stands for degrees of freedom for groups and DFE stands for degrees of freedom for error. DFG = I -1, the number of groups minus 1. DFE = N–I, the number of observations minus the number of groups. The P-value for this distribution is also given. In this example, the P-value is given as 0.000. This means that the P-value is less than 0.001. This is strong evidence that the means are not all equal.

For information purposes, the output from one-way analysis of variance provides the mean and standard deviation for each group and plots individual 95% confidence intervals for the means. Each confidence interval is of the form

$$\left(\bar{x}_i - t^* \frac{s_p}{\sqrt{n_i}}, \bar{x}_i + t^* \frac{s_p}{\sqrt{n_i}}\right)$$

where $\bar{x}_i$ and n_i are the sample mean and sample size for level i, s_p= Pooled StDev is the pooled estimate of the common standard deviation, and t^* is the value from a t table corresponding to 95% confidence and the degrees of freedom associated with MS Error.

EXERCISES

22.3 How does logging in a tropical rainforest affect the forest several years later? Researchers compared forest plots in Borneo that had never been logged (Group 1) with similar plots nearby that had been logged 1 year earlier (Group 2) and 8 years earlier (Group 3). Although the study was not an experiment, the authors explain why we can consider the plots to be randomly selected. The data appear in Table 22.2 of BPS and TA22-02.MTw. The variable Trees is the count of trees in a plot; Species is the count of tree species in a plot. The variable Richness is the ratio of number of species to number of individual trees, calculated as Species/Trees. Select **Stat ➤ ANOVA ➤ One-Way** from the menu to analyze the data.

(a) Click on the Graphs button to make side-by-side stemplots of Trees for the three groups. What effects of logging are visible?

(b) What do the group means show about the effects of logging?

(c) What are the values of the ANOVA F statistic and its P-value? What hypotheses does F test? What conclusions about the effects of logging on number of trees do the data lead to?

22.4 If you are a dog lover, perhaps having your dog along reduces the effect of stress. To examine the effect of pets in stressful situations, researchers recruited 45 women who said they were dog lovers. The EESEE story "Stress among Pets and Friends" describes the results. Fifteen of the subjects were randomly assigned to each of three groups to do a stressful task alone (the control group), with a good friend present, or with their dog present. The subject's mean heart rate during the task is one measure of the effect of stress. Table 22.3 and TA 22-03.MTW contains the data. Select **Stat ➤ ANOVA ➤ One-Way** from the menu to analyze these data.

(a) Select **Graph ➤ Stem-and-Leaf** from the menu to make stemplots of the heart rates for the three groups (round to the nearest whole number of beats). Do any of the groups show outliers or extreme skewness?

(b) Do the mean heart rates for the groups appear to show that the presence of a pet or a friend reduces heart rate during a stressful task?

(c) What are the values of the ANOVA F statistic and its P-value? What hypotheses does F test? Briefly describe the conclusions you draw from these data. Did you find anything surprising?

22.9 Table 22.2 in BPS and TA22-02.MTW give data on the species richness in rainforest plots, defined as the number of tree species in a plot divided by the number of trees in the plot. ANOVA may not be trustworthy for the richness data. Select **Graph ➤ Stem-and-Leaf** from the menu to make stemplots to examine the distributions of the response variable in the three groups and also compare the standard deviations. What about the data makes ANOVA risky?

22.22 How quickly do synthetic fabrics such as polyester decay in landfills? A researcher buried polyester strips in the soil for different lengths of time, then dug up the strips and measured the force required to break them. Breaking strength is easy to measure and is a good indicator of decay; lower strength means the fabric has decayed. Part of the study buried 20 polyester strips in well-drained soil in the summer. Five of the strips, chosen at random, were dug up after 2 weeks; another 5 were dug up after 4 weeks, 8 weeks, and 16 weeks. Here and in EX22-22.MTW are the breaking strengths in pounds. Select **Stat ➤ Basic Statistics ➤ Display Descriptive Statistics** to obtain the means and standard deviations for these data.

2 weeks	118	126	126	120	129
4 weeks	130	120	114	126	128
8 weeks	122	136	128	146	140
16 weeks	124	98	110	140	110

(a) Does it appear that polyester loses strength consistently over time after it is buried?

(b) Do the standard deviations meet our criterion for applying ANOVA?

(c) Explain carefully why using the *F* test on all four groups is not acceptable.

22.25 How much corn per acre should a farmer plant to obtain the highest yield? Too few plants will give a low yield. On the other hand, if there are too many plants, they will compete with each other for moisture and nutrients, and yields will fall. To find out, plant at different rates on several plots of ground and measure the harvest. (Treat all the plots the same except for the planting rate.) Select **Stat ➤ ANOVA ➤ One-Way** from the menu to analyze these data from such an experiment:

Plants per acre	Yield (bushels per acre)			
12,000	150.1	113.0	118.4	142.6
16,000	166.9	120.7	135.2	149.8
20,000	165.3	130.1	139.6	149.9
24,000	134.7	138.4	156.1	
28,000	119.0	150.5		

(a) Do data analysis to see what the data appear to show about the influence of plants per acre on yield and also to check the conditions for ANOVA.

(b) Carry out the ANOVA *F* test. State hypotheses and give *F* and its *P*-value. What do you conclude?

(c) The observed differences among the mean yields in the sample are quite large. Why are they not statistically significant?

22.26 Table 22.2 in BPS and TA22.-2.MTW give data on the number of trees per forest plot, the number of species per plot, and species richness. Exercise 22.3 analyzed the effect of logging on number of trees. Exercise 22.9 concluded that it would be risky to use ANOVA to analyze richness. Select **Stat ➤ ANOVA ➤ One-Way** to analyze the effect of logging on the number of species.

(a) Do the standard deviations satisfy our rule of thumb for safe use of ANOVA? What do the means suggest about the effect of logging on the number of species?

(b) Report the *F* statistic and its *P*-value and state your conclusion.

22.27 A botanist prepares 16 identical planting pots and then introduces different numbers of nematodes into the pots. He transplants a tomato seedling into each plot. Here and in EX22-27.MTW are data on the increase in height of the seedlings (in centimeters) 16 days after planting:

Nematodes	Seedling growth			
0	10.8	9.1	13.5	9.2
1,000	11.1	11.1	8.2	11.3
5,000	5.4	4.6	7.4	5.0
10,000	5.8	5.3	3.2	7.5

(a) Select **Stat ➤ Basic Statistics ➤ Display Descriptive Statistics** to obtain the means and standard deviations for the four treatments. Select **Graph ➤ Boxplots** to make side-by-side stemplots to compare the treatments. What do the data appear to show about the effect of nematodes on growth?

(b) State H_0 and H_a for the ANOVA test for these data, and explain in words what ANOVA tests in this setting.

(c) Select **Stat ➤ ANOVA ➤ One-Way** to carry out the ANOVA. Report your overall conclusions about the effect of nematodes on plant growth.

Chapter 23
Nonparametric Tests

Topic to be covered in this chapter:

Wilcoxon Rank Sum Test
Wilcoxon Signed Rank Test
Kruskal-Wallis Test

Wilcoxon Rank Sum Test

Minitab does a two-sample rank test (often called the Mann-Whitney test or the two-sample Wilcoxon rank sum test) for the difference between two population medians, and calculates the corresponding point estimate and confidence interval. To perform the Wilcoxon rank sum test select

Stat ➤ Nonparametrics ➤ Mann-Whitney

from the menu.

Example 23.1 describes the following setting for the Wilcoxon rank sum test. A researcher planted corn in 8 plots of ground, then weeded the corn to allow no weeds in 4 plots and exactly 3 weeds per meter in the other 4 plots. The table here and in EX23-01.MTW shows yields of corn (bushels per acre) in each of the plots.

Weeds per meter	Yield (bu/acre)			
0	166.7	172.2	165.0	176.9
3	158.6	176.4	153.1	156.0

Minitab assumes that the data are independent random samples from two populations that have the same shape (hence the same variance) and a scale that is at least ordinal. The data need not be from normal populations. The data for this command needs to be in two columns containing data from two populations. The columns do not need to be the same length.

Since the data in TA23-01.MTW are arranged with the number of weeds in one column and the yield in another, it is necessary to unstack the data in two columns. To unstack the data, select

Data ➤ Unstack Columns

from menu. In the dialog box, enter "Yield" after Unstack the data in and "Weeds" after Using subscripts in. Store the unstacked data after the last column in use.

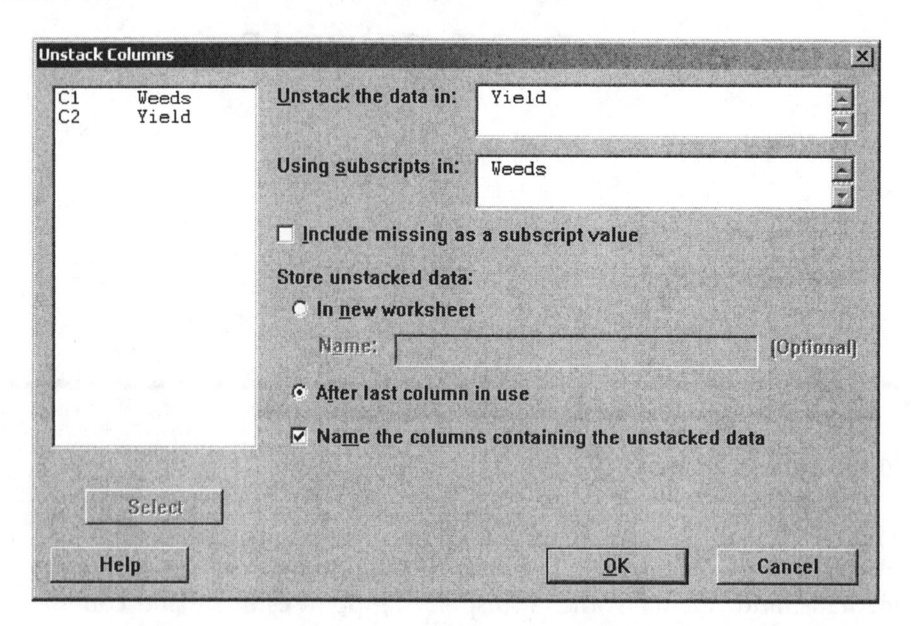

Once the new columns are added, the worksheet will contain the following data.

↓	C1	C2	C3	C4	C5	C6
	Weeds	Yield	Yield_0	Yield_3		
1	0	166.7	166.7	158.6		
2	0	172.2	172.2	176.4		
3	0	165.0	165.0	153.1		
4	0	176.9	176.9	156.0		
5	3	158.6				
6	3	176.4				
7	3	153.1				
8	3	156.0				

Perform the test by selecting **Stat ➤ Nonparametrics ➤ Mann-Whitney** from the menu. In the First Sample and Second Sample columns, enter the columns with sample data, select the appropriate Alternative, and click OK. The Mann-Whitney tests H_0: $median_1 = median_2$ against a one- or two-sided alternative. Since we expect a higher yield in the weed-free plots, we select the "greater than" alternative.

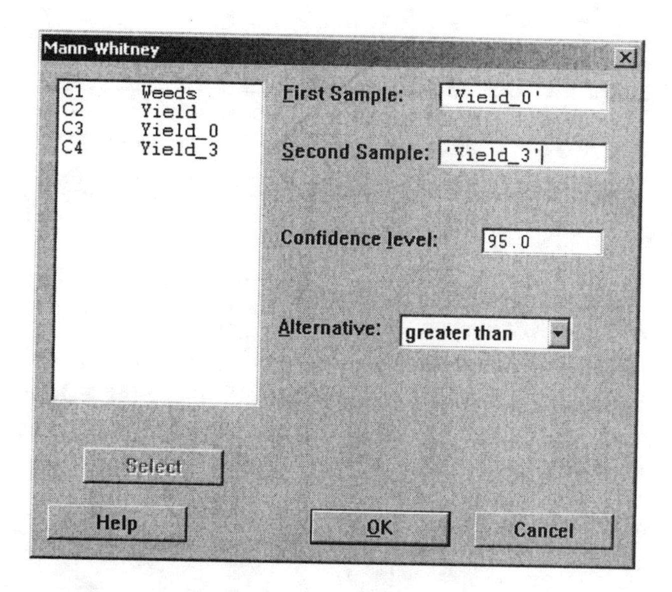

Minitab determines the attained significance level of the test using a normal approximation with a continuity correction factor. We see from the output that follows that the sum of the ranks in the first group (0 weeds) is W = 23, with approximate *P*-value 0.0970. The effect of weeds on yield is not statistically significant at the 0.05 level.

Mann-Whitney Test and CI: Yield_0, Yield_3

```
         N  Median
Yield_0  4  169.45
Yield_3  4  157.30


Point estimate for ETA1-ETA2 is 11.30
97.0 Percent CI for ETA1-ETA2 is (-11.40,23.80)
W = 23.0
Test of ETA1 = ETA2 vs ETA1 > ETA2 is significant at 0.0970
```

Wilcoxon Signed Rank Test

Minitab performs a one-sample Wilcoxon signed rank test of the median for single samples or matched pairs. To perform the Wilcoxon signed rank test, select

Stat ➤ Nonparametrics ➤ 1-Sample Wilcoxon

from the menu. The Wilcoxon test assumes that the data are a random sample from a symmetric population that is not necessarily normal.

Example 23.6 in BPS considers a study of early childhood education. Kindergarten students were asked to tell a fairy tale that had been read to them earlier in the week. Each child told two stories. The first had been read to them and the second had been read and also illustrated with pictures. An expert listened to a recording of the children and assigned a score for certain uses of language. Here and in EG23-06.MTW are the data for five "low progress" readers in a pilot study:

Child	1	2	3	4	5
Story 1	0.77	0.49	0.66	0.28	0.38
Story 2	0.40	0.72	0.00	0.36	0.55
Difference	0.37	−0.23	0.66	−0.08	−0.17

We will test the hypotheses

H_0: scores have the same distribution for both stories

H_a: scores are systematically higher for Story 2

Because these are matched pairs data, we base our inference on the differences. Enter the differences into a Minitab worksheet. Select **Stat ➤ Nonparametrics ➤ 1-Sample Wilcoxon** from the menu. Enter the variable and specify the null hypothesis and the alternative, as shown in the dialog box.

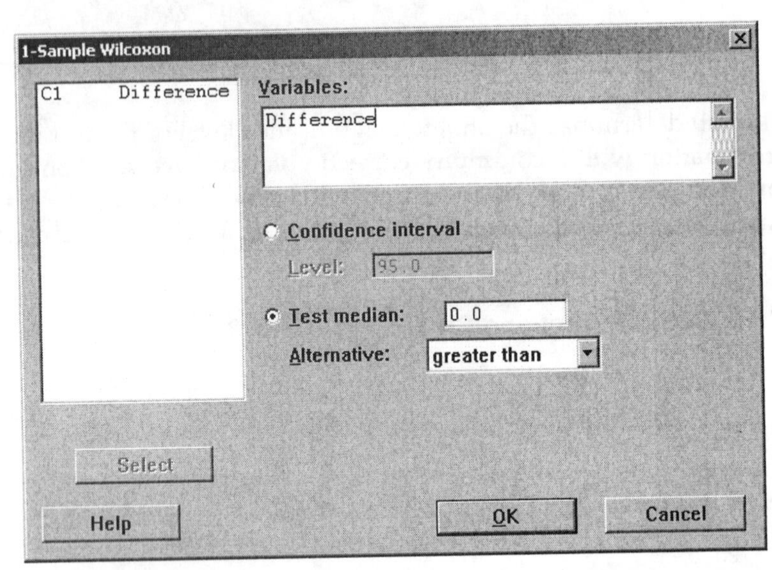

The output following shows that the observed value $W^+ = 9$. Minitab uses the Normal approximation with the continuity correction to give the approximate P-value of 0.394. This small sample is not statistically significant.

Wilcoxon Signed Rank Test: Difference

```
Test of median = 0.000000 versus median > 0.000000

                  N
                 for   Wilcoxon                Estimated
              N  Test  Statistic        P         Median
Difference    5     5        9.0    0.394         0.1000
```

Kruskal-Wallis Test

The Kruskal-Wallis test is a rank test that can replace the ANOVA F test. This test is a generalization of the procedure used by Mann-Whitney. The test assumes that the data arise as k independent random samples from continuous distributions, all having the same shape (normal or not). The null hypothesis of no differences among the k populations is tested against the alternative of at least one difference. To perform the test, select

Stat ➤ Nonparametrics ➤ Kruskal-Wallis

from the menu. The factor column may be numeric or text, and may contain any value. The levels do not need to be in any special order.

Example 23.11 in BPS provides an illustration of the Kruskal-Wallis test. The data in the example and EG23-11.MTW gives the number of weeks allowed to grow in each plot and the corn yield. The hypotheses tested in the example are

H_0: yields have the same distribution in all groups

H_a: yields are systematically higher in some groups than in others

The Kruskal-Wallis test can be performed by selecting **Stat ➤ Nonparametrics ➤ Kruskal-Wallis** from the menu and filling in the columns for Response and Factor as shown.

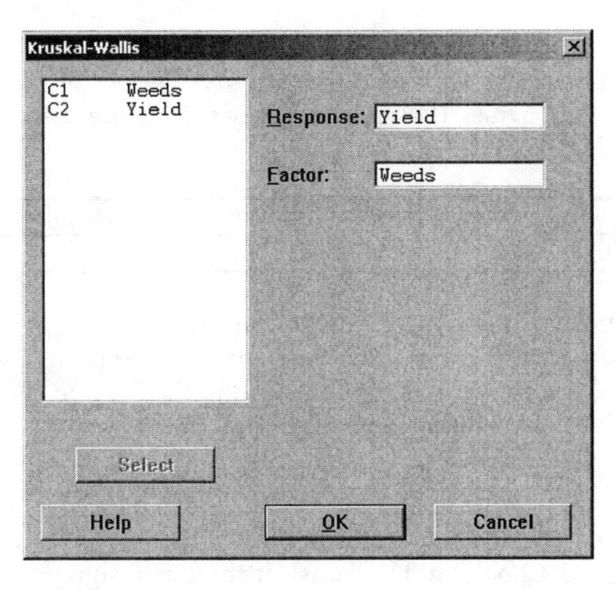

The following output gives the results $H = 5.56$ and $P = 0.135$, which do not provide convincing evidence that weeds have an effect on yield.

Kruskal-Wallis Test: Yield versus Weeds

```
Kruskal-Wallis Test on Yield

Weeds      N  Median  Ave Rank      Z
0          4   169.5      13.1   2.24
1          4   163.7       8.4  -0.06
3          4   157.3       6.3  -1.09
9          4   162.6       6.3  -1.09
Overall   16               8.5

H = 5.56  DF = 3  P = 0.135
H = 5.57  DF = 3  P = 0.134  (adjusted for ties)

* NOTE * One or more small samples
```

EXERCISES

23.1 To detect the presence of harmful insects in farm fields, we can put up boards covered with a sticky material and examine the insects trapped on the boards. Which colors attract insects best? Experimenters placed boards of several colors at random locations in a field of oats. Here and in EX23-01.MTW are the counts of cereal leaf beetles trapped by boards colored blue and green:

Blue	16	11	20	21	14	7
Green	37	32	20	29	37	32

Because the samples are small, it is difficult to verify approximate Normality. Select **Data ➤ Unstack Columns** to unstack the data. Then select **Stat ➤ Nonparametrics ➤ Mann-Whitney** from the menu to use the Wilcoxon rank sum test. State hypotheses, and give the *P*-value and your conclusions.

23.2 Exercise 17.7 describes an experiment that compared the change in the level of polyphenols in the blood after two weeks of drinking either red wine or white wine. (Polyphenols may reduce the risk of a heart attack.) Here and in EX23-02.MTW are the data:

Red wine	3.5	8.1	7.4	4.0	0.7	4.9	8.4	7.0	5.5
White wine	3.1	0.5	−3.8	4.1	−0.6	2.7	1.9	−5.9	0.1

Select **Data ➤ Unstack Columns** to unstack the data. Then select **Stat ➤ Nonparametrics ➤ Mann-Whitney** from the menu to use the Wilcoxon rank sum test. State your hypotheses and the *P*-value. What do you conclude about red wine versus white wine?

23.8 Exercise I.3 in BPS and EX0I-03.MTW contain data from a sample of first-year college students who were asked how long (in minutes) they study on a typical school night. We ask if there is a significant difference between men and women.

(a) Select **Stat ➤ Basic Statistics ➤ 2-Sample t** to carry out the two-sample t test. What are t and its two-sided *P*-value?

(b) Select **Data ➤ Unstack Columns** to unstack the data. Then select **Stat ➤ Nonparametrics ➤ Mann-Whitney** from the menu to use the Wilcoxon rank sum test. What are W and its two-sided *P*-value. Do t and W lead to the same practical conclusion?

(c) One male student in the sample claimed to study 30,000 minutes per night. This was clearly a joke, so this student was replaced by another who claimed 180 minutes. Replace the final 180 in the male data by 30,000. Recalculate t and W and their *P*-values. Compare the effect of the outlier on the two tests.

23.15 Table 17.1 in BPS and TA17-01.MTW give the pretest and posttest scores for two groups of students taking a program to improve their basic mathematics skills. Did the treatment group show significantly greater improvement than the control group?

(a) Select **Calc ➤ Calculator** to calculate the post-test versus pre-test differences. Then select **Data ➤ Unstack Columns** to unstack the data into separate columns for the two groups of students. Finally, select **Stat ➤ Nonparametrics ➤ Mann-Whitney** from the menu to apply the Wilcoxon rank sum test to the posttest versus pretest differences. What do you conclude?

(b) Select **Stat ➤ Basic Statistics ➤ 2-Sample t** to compare your findings with those from the two-sample t test.

(c) What are the null and alternative hypotheses for each of the two tests we have applied to these data?

(d) What must we assume about the data to apply each of the tests?

23.16 Example 23.4 in BPS describes a study of the attitudes of people attending outdoor fairs about the safety of the food served at such locations. The full data set is given in EX23-16.MTW. It contains the responses of 303 people to several questions. The variables in this data set are (in order): subject, hfair, sfair, sfast, srest, and gender. The variable sfair contains the responses described in the example concerning safety of food served at outdoor fairs and festivals. The variable srest contains responses to the same question asked about food served in restaurants. The variable gender contains 1 if the respondent is a woman, 2 if the respondent is a man. We saw that women are more concerned than men about the safety of food served at fairs. Select **Data ➤ Unstack Columns** to unstack the data into separate columns for men and women. Then select **Stat ➤ Nonparametrics ➤ Mann-Whitney** from the menu to apply the Wilcoxon rank sum test to see if this is also true for restaurants.

23.20 Table 16.1 in BPS and EX23.20 gives matched pairs data for 21 subjects. The response variable is time to complete a maze, both wearing a scented mask and wearing an identical mask that is unscented. Does the scent improve performance (that is, shorten the time needed to complete the maze)? The matched pairs t test (Example 16.3 in BPS) works well, and gives $P = 0.365$. Select **Stat ➤ Nonparametrics ➤ 1-Sample Wilcoxon** from the menu to compare the Wilcoxon signed rank test. What is the value of W^+? Find the P-value for the test. Does the Wilcoxon signed rank test lead to essentially the same result as the t test?

23.24 Exercise 16.7 of BPS reports the following data on the percent of nitrogen in bubbles of ancient air trapped in amber:

63.4 65.0 64.4 63.3 54.8 64.5 60.8 49.1 51.0

We wonder if ancient air differs significantly from the present atmosphere, which is 78.1% nitrogen.

(a) Graph the data, and comment on skewness and outliers. A rank test is appropriate.

(b) We would like to test hypotheses about the median percent of nitrogen in ancient air (the population):

$$H_0\text{: median} = 78.1$$

$$H_a\text{: median} \neq 78.1$$

Select **Stat ➤ Nonparametrics ➤ 1-Sample Wilcoxon** from the menu to apply the Wilcoxon signed rank test to the differences between the observations and 78.1. (This is the one-sample version of the test.) What do you conclude?

23.25 The EESEE story "Stepping Up Your Heart EESEE Rate" describes a student project that asked subjects to step up and down for three minutes and measured their heart rates before and after the exercise. Here and in EX23-25.MTW are data for five subjects and two treatments: stepping at a low rate (14 steps per minute) and at a medium rate (21 steps per minute). For each subject, we give the resting heart rate (beats per minutes) and the heart rate at the end of the exercise. (The data are slightly modified.)

	Low Rate		Medium Rate	
Subject	Resting	Final	Resting	Final
1	60	75	63	84
2	90	98	69	93
3	87	93	81	96
4	78	87	75	90
5	84	84	90	108

Does exercise at the low rate raise heart rate significantly?

(a) State hypotheses in terms of the median increase in heart rate.

(b) Select **Stat ➤ Nonparametrics ➤ 1-Sample Wilcoxon** from the menu to apply the Wilcoxon signed rank test. Do the data give good evidence that stepping at the medium rate raises heart rates? What is the value of W^+? Find the *P*-value and state your conclusion.

23.28 Table 16.3 in BPS and EX23-28.MTW give data on the healing rate (micrometers per hour) of the skin of newts under two conditions. This is a matched pairs design, with the body's natural electric field for one limb (control) and half the natural value for another limb of the same newt (experimental). We want to know if the healing rates are systematically different under the two conditions. You decide to use a rank test. Select **Stat ➤ Nonparametrics ➤ 1-Sample Wilcoxon** to find the signed rank statistic W^+ and the *P*-value. Give a conclusion. Be sure to include a description of what the data show in addition to the test results.

23.29 Cola makers test new recipes for loss of sweetness during storage. Trained tasters rate the sweetness before and after storage. Here and in EX23-29.MTW are the sweetness losses (sweetness before storage minus sweetness after storage) found by 10 tasters for one new cola recipe:

2.0 0.4 0.7 2.0 −0.4 2.2 −1.3 1.2 1.1 2.3

Are these data good evidence that the cola lost sweetness?

(a) These data are the differences from a matched pairs design. State hypotheses in terms of the median difference in the population of all tasters. Select **Stat ➤ Nonparametrics ➤ 1-Sample Wilcoxon** to carry out a test. Give your conclusion.

(b) Select **Stat ➤ Basics Statistics ➤ 1-Sample t** from the menu to find the one-sample t test P-value for these data. How does this compare with your result from (a)? What are the hypotheses for the t test? What assumptions must we make for each of the t and Wilcoxon tests?

23.30 Example 22.6 in BPS used ANOVA to analyze the results of a study to see which of four colors best attracts cereal leaf beetles. Here and in EX23-30.MTW are the data:

Color	Insects trapped					
Lemon yellow	45	59	48	46	38	47
White	21	12	14	17	13	17
Green	37	32	15	25	39	41
Blue	16	11	20	21	14	7

Because the samples are small, we will apply a nonparametric test. Select **Stat ➤ Nonparametrics ➤ Kruskal-Wallis** from the menu to calculate the Kruskal-Wallis statistic H. What does the test lead you to conclude?

23.31 Table 22.2 in BPS and EX23-31.MTW contains data comparing the number of trees and number of tree species in plots of land in a tropical rainforest that had never been logged with similar plots nearby that had been logged 1 year earlier and 8 years earlier. The third response variable is species richness, the number of tree species divided by the number of trees. There are low outliers in the data, and a histogram of the ANOVA residuals shows outliers as well. Because of lack of Normality in small samples, we may prefer the Kruskal-Wallis test.

(a) Make a graph to compare the distributions of richness for the three groups of plots. Also give the median richness for the three groups.

(b) Select **Stat ➤ Nonparametrics ➤ Kruskal-Wallis** from the menu to use the Kruskal-Wallis test to compare the distributions of richness. State hypotheses, the test statistic and its P-value, and your conclusions.

23.32 Here are the breaking strengths (in pounds) of strips of polyester fabric buried in the ground for several lengths of time:

2 weeks	118	126	126	120	129
4 weeks	130	120	114	126	128
8 weeks	122	136	128	146	140
16 weeks	124	98	110	140	110

Breaking strength is a good measure of the extent to which the fabric has decayed. The standard deviations of the 4 samples do not meet our rule of thumb for applying ANOVA. In addition, the sample buried for 16 weeks contains an outlier. Select **Stat ➤ Nonparametrics ➤ Kruskal-Wallis** from the menu to use the nonparametric test. What are the hypotheses for the Kruskal-Wallis test, expressed in terms of medians? Report your conclusion.

23.34 Exercise 17.13 in BPS reports the results of a study of the effect of the pesticide DDT on nerve activity in rats. The data for the DDT group are in EX23-34.MTW and below

12.207 16.869 25.050 22.429 8.456 20.589

The control group data are

11.074 9.686 12.064 9.351 8.182 6.642

It is difficult to assess Normality from such small samples, so we might use a nonparametric test to assess whether DDT affects nerve response. Select **Data ➤ Unstack Columns** to unstack the data. Then select **Stat ➤ Nonparametrics ➤ Mann-Whitney** from the menu to use the Wilcoxon rank sum test. Compare the responses in the two groups. State hypotheses and give your conclusions.

23.36 Exercise 17.8 compares the number of tree species in unlogged plots in the rain forest of Borneo with the number of species in plots logged 8 years earlier. Here and in EX23-36.MTW are the data:

Unlogged	22	18	22	20	15	21	13	13	19	13	19	15
Logged	17	4	18	14	18	15	15	10	12			

(a) Select **Graph ➤ Stem-and-Leaf** to make stemplots of the data. Does there appear to be a difference in species counts for logged and unlogged plots?

(b) Does logging significantly reduce the mean number of species in a plot after 8 years? State the hypotheses. Select **Data ➤ Unstack Columns** to unstack the data. Then select **Stat ➤ Nonparametrics ➤ Mann-Whitney** from the menu to do a rank test, and state your conclusion.

23.39 A botanist prepares 16 identical planting pots and then introduces different numbers of nematodes into the pots. A tomato seedling is transplanted into each plot. Here and in EX23.39.MTW are data on the increase in height of the seedlings (in centimeters) 16 days after planting.

Nematodes	Seedling growth			
0	10.8	9.1	13.5	9.2
1,000	11.1	11.1	8.2	11.3
5,000	5.4	4.6	7.4	5.0
10,000	5.8	5.3	3.2	7.5

We applied ANOVA to these data in Exercise 22.27. Because the samples are very small, it is difficult to assess Normality.

(a) What hypotheses does ANOVA test? What hypotheses does Kruskal-Wallis test?

(b) Do nematodes appear to retard growth? Select **Stat ➤ Nonparametrics ➤ Kruskal-Wallis** from the menu to apply the Kruskal-Wallis test. What do you conclude?

Chapter 24
Statistical Process Control

Topic to be covered in this chapter:

Pareto Charts
Control Charts for Sample Means
Out-of-Control Signals
Control Charts for Sample Proportions

Pareto Charts

Pareto charts are bar graphs with the bars ordered by height. They are often used to isolate the "vital few" categories on which we should focus our attention. Exercise 24.4 of BPS describes the following example. A large medical center, financially pressed by restrictions on reimbursement by insurers and the government, looked at losses broken down by diagnosis. Government standards place cases into diagnostic related groups (DRGs). For example, major joint replacements (mostly hip and knee) are DRG 209. The data in Exercise 24.4 and EX24-04.MTW list the 9 DRGs with the most losses along with the percent of losses. Since the percents are given, a Pareto chart can be constructed by selecting **Graph ➤ Bar Chart** from the menu. Indicate that the bars represent "Values from a table" and select Simple.

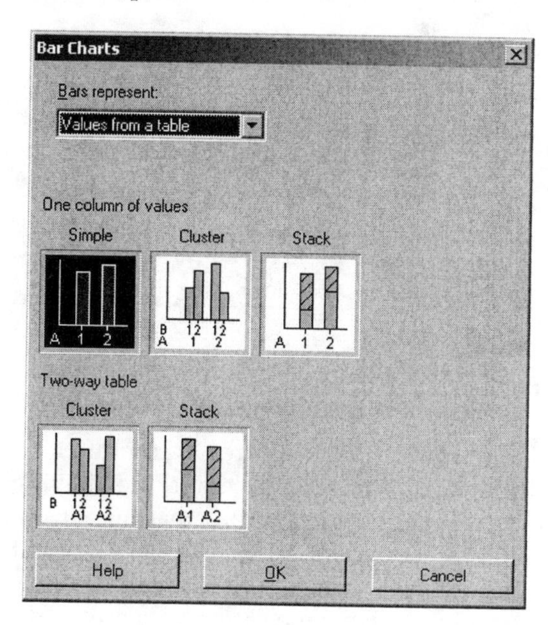

In the following dialog box, indicate that the graph variable is 'Percent of Losses' and the categorical variable is DRG.

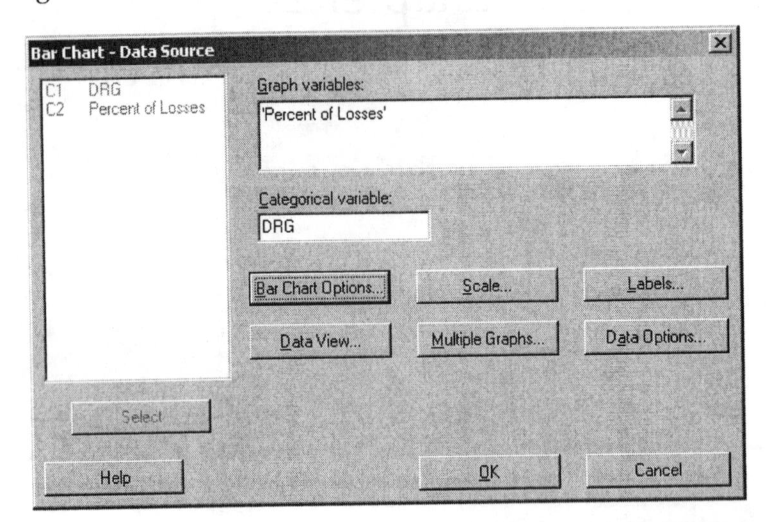

Click on the Bar Chart Options button and choose order based on Decreasing Y in the Bar Chart Options subdialog box. Click OK on the dialog boxes to make the Pareto chart. The Pareto chart helps the hospital decide which DRGs the hospital should study first when attempting to reduce its losses.

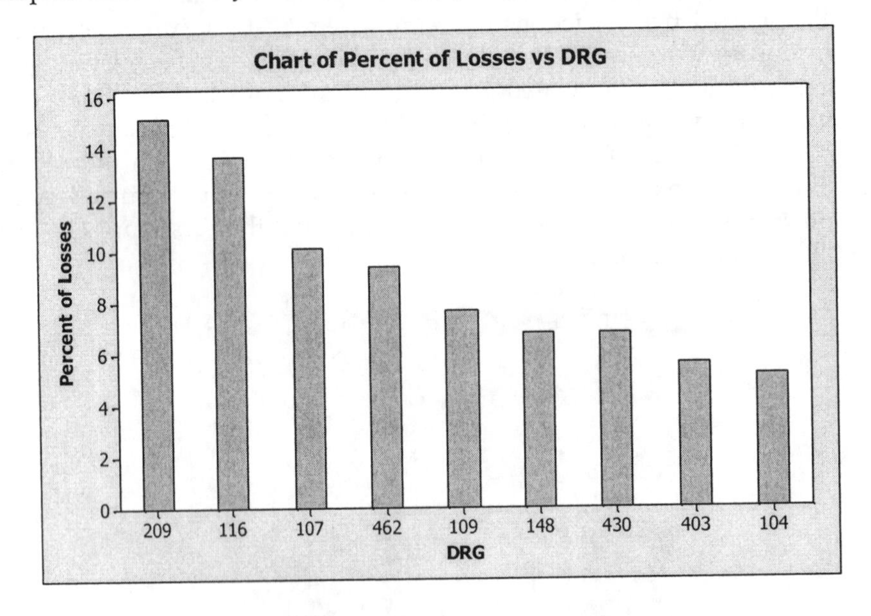

Control Charts for Sample Means

Example 24.3 of BPS discusses a manufacturer of computer monitors. The manufacturer measures the tension of fine wires behind the viewing screen. Tension is measured by an electrical device with output readings in millivolts (mV). The proper tension is 275 mV. Some variation is always present in the production process. When the process is operating properly, the standard deviation of the tension readings is $\sigma = 43$ mV. Four measurements are made every hour. Table

24.1 of BPS and TA24-01.MTW contain the measurements for 20 hours. The first row of observations is from the first hour, the next row is from the second hour, and so on. There are a total of 80 observations.

TA24-01.MTW ***

↓	C1 Sample	C2 s1	C3 s2	C4 s3	C5 s4	C6 x-bar	C7 s
1	1	234.5	272.3	234.5	272.3	253.4	21.8
2	2	311.1	305.8	238.5	286.2	285.4	33.0
3	3	247.1	205.3	252.6	316.1	255.3	45.7
4	4	215.4	296.8	274.2	256.8	260.8	34.4
5	5	327.9	247.2	283.3	232.6	272.7	42.5
6	6	304.3	236.3	201.8	238.5	245.2	42.8
7	7	268.9	276.2	275.6	240.2	265.2	17.0
8	8	282.1	247.7	259.8	272.8	265.6	15.0
9	9	260.8	259.9	247.9	345.3	278.5	44.9
10	10	329.3	231.8	307.2	273.4	285.4	42.5
11	11	266.4	249.7	231.5	265.2	253.2	16.3
12	12	168.8	330.9	333.6	318.3	287.9	79.7

Minitab can be used to produce control charts for sample means by selecting

Stat ➤ Control Charts ➤ Variable Chart for Subgroups ➤ Xbar

from the menu. An $\bar{x}$-control chart can be made using either the raw sample data in columns C2–C5 or the $\bar{x}$ data. To make a control chart using the $\bar{x}$ data, specify that the data are arranged as a Single column and select the column with the $\bar{x}$ data. You must also specify that the Subgroup size is 1, as shown in the dialog box.

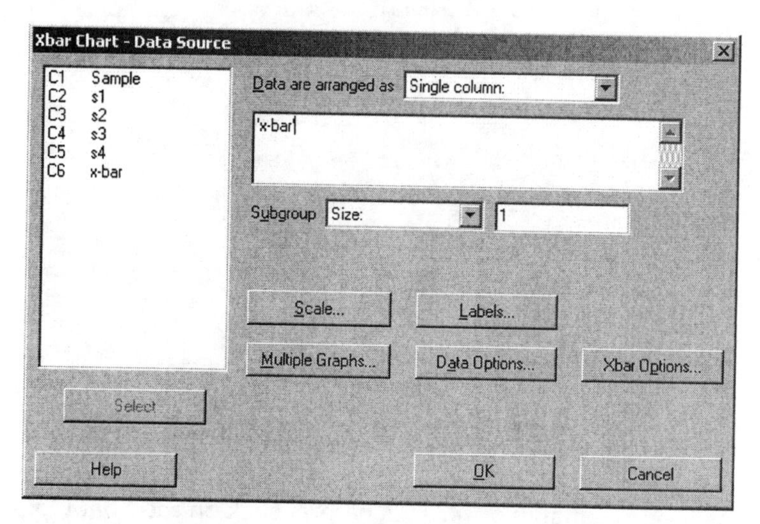

The $\bar{x}$ values will be plotted on the chart. In addition, a center line, an upper control limit (UCL) at 3σ above the center line, and a lower control limit (LCL) at 3σ below the center line are drawn on the chart. By default, the process mean μ and standard deviation σ are estimated from the data. Alternatively, the parameters μ and σ may be specified from historical data by clicking on the Xbar Options button and selecting the Parameters tab. In the following dialog box we specify that the historical mean is equal to 275 and the historical standard deviation is equal to $\sigma/\sqrt{n} = 21.5$.

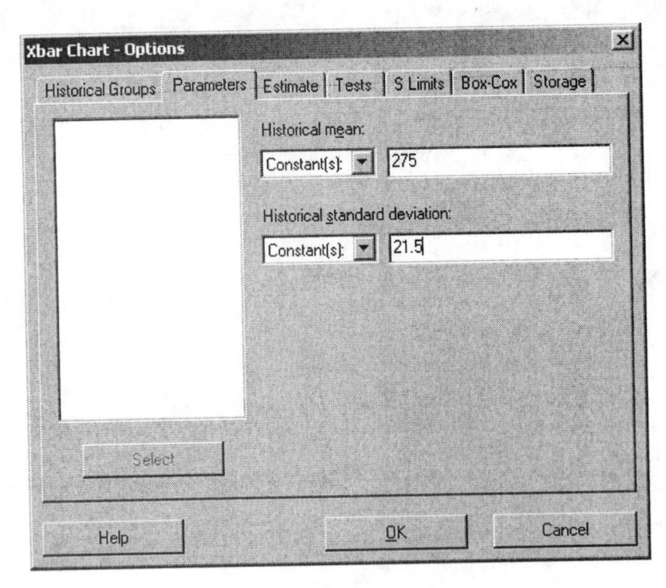

Alternatively, the $\bar{x}$-control chart can be made by specifying that the data are arranged as "Subgroups of rows of" and select the columns containing the raw data.

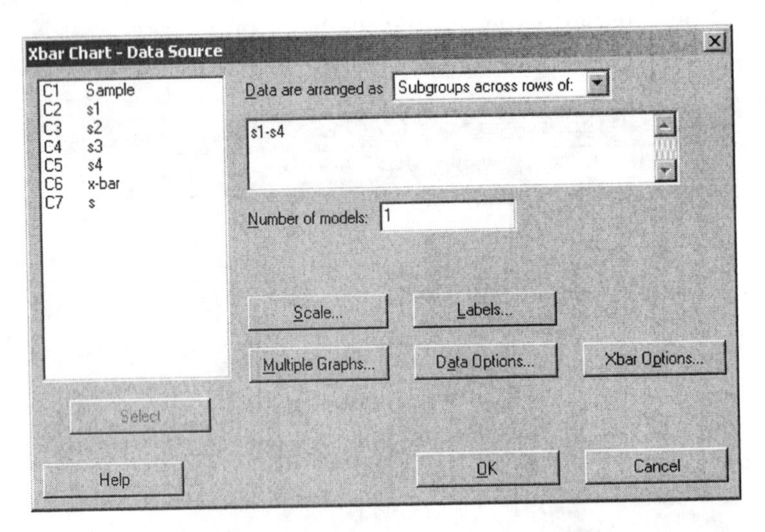

In this case, the historical mean is still equal to 275, but the historical standard deviation is equal to $\sigma = 43$. As before, click on the Xbar Options button and the Parameters tab. Both methods will produce an $\bar{x}$-control chart (with a slightly different title), as follows.

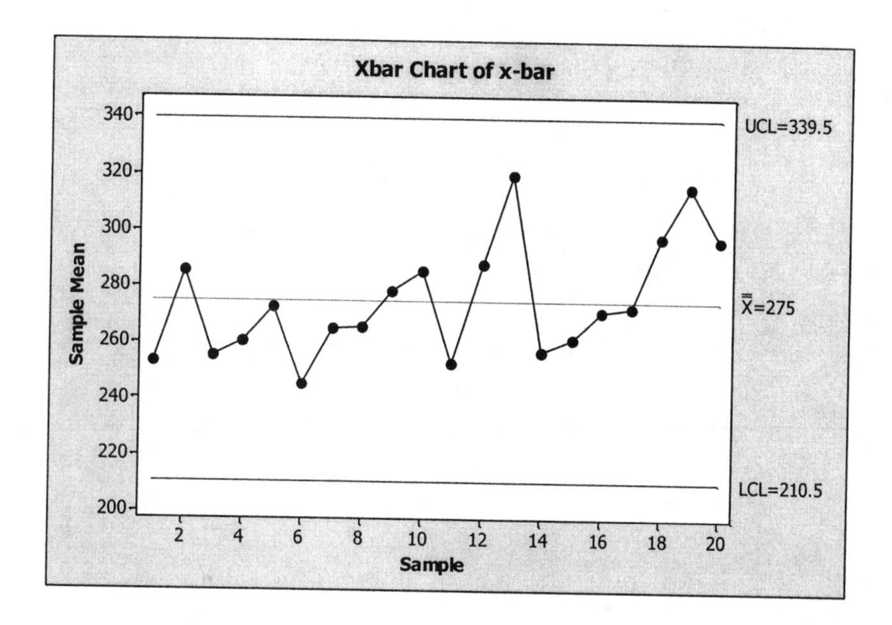

In this $\bar{x}$ chart, no points lie outside the control limits. In practice, we must monitor both the process center, using an $\bar{x}$ chart, and the process spread, using a control chart for the sample standard deviation s.

In practice, we must control the center of a process and its variability. This is commonly done with an s chart, a chart of standard deviations against time. The s chart can be produced by selecting

Stat ➤ Control Charts ➤ Variables Charts for Subgroups ➤ S

from the menu. The dialog box is filled exactly the same way as the dialog box for the $\bar{x}$ chart on the previous page. Often, s minus three standard deviations gives a negative number. In this case, the lower control limit is plotted at 0.

The s chart for the mesh tension data is also in control. Usually, the $\bar{x}$ chart and the s chart will be looked at together. We can produce both charts at once by selecting

Stat ➤ Control Charts ➤ Variables Charts for Subgrous ➤ Xbar-S

from the menu. As usual, we have a choice of specifying or not specifying the historical values of μ and σ. These affect the centerline and control limits of the $\bar{x}$ chart only.

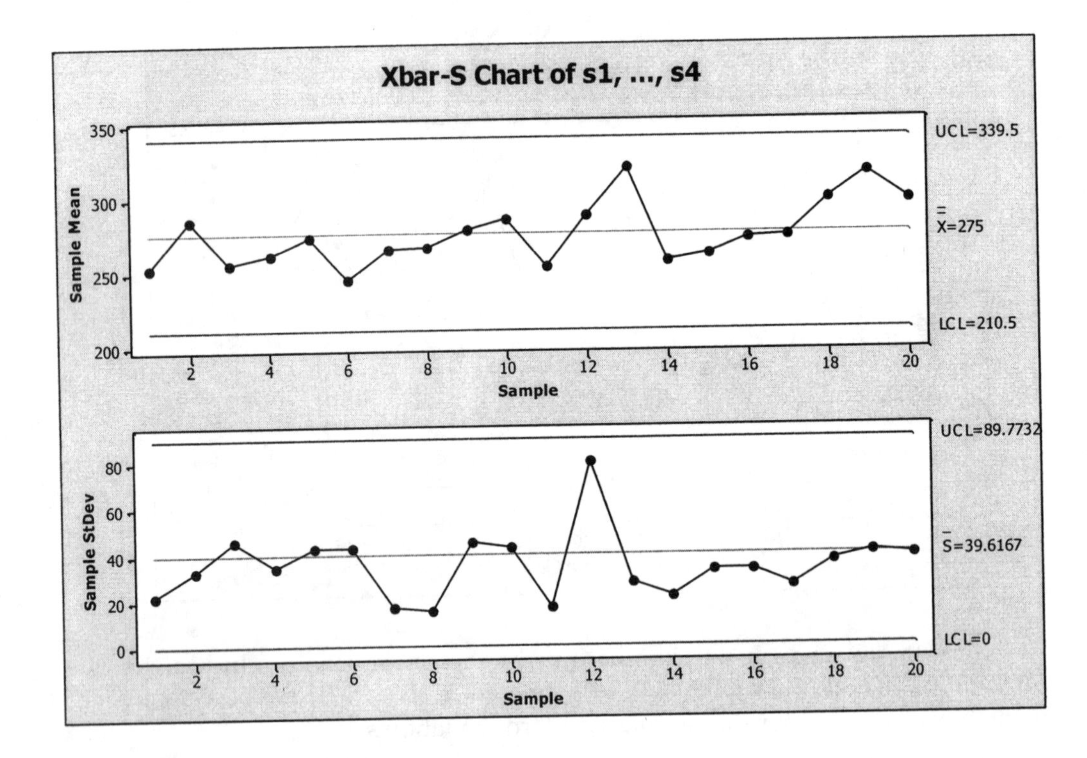

Out-of-Control Signals

Minitab performs tests to identify out-of- control signals. Each test detects a specific pattern in the data plotted on the chart. The occurrence of a pattern suggests a special cause for the variation, one that should be investigated. The tests can be selected by clicking the Xbar Options button on the dialog box and choosing from the Tests tab.

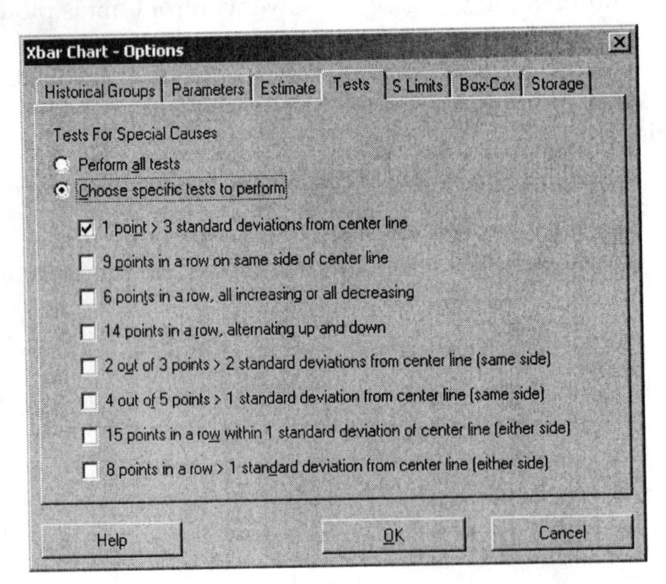

When a point fails a test, it is marked with the test number on the plot. If a point fails more than one test, the number of the first test in your list is the number displayed on the plot.

Control Charts for Sample Proportions

Example 24.13 of BPS discusses using p charts for manufacturing and school absenteeism. Table 24.8 of BPS and TA24-08.MTW contain data on production workers and record the number and proportion absent from work each day during a period of four weeks. Minitab can be used to produce control charts for proportions by selecting

Stat ➤ Control Charts ➤ Attributes chart ➤ P

from the menu. Minitab draws a p chart to show the proportion absent (the number absent divided by the subgroup size). P charts track the proportion absent and detects the presence of special causes. Each entry in the worksheet column is the number absent for one subgroup. In the dialog box under Variables, enter the column that contain the number absent (or defective) for each sample. Choose "Size" for subgroups, then enter the subgroup size as shown.

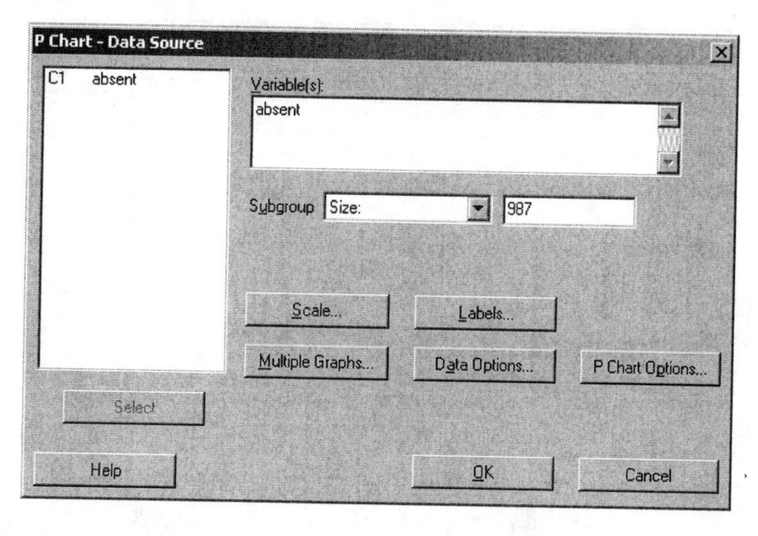

Choose 'ID column' for unequal-size subgroups, then enter a column of subscripts. If the subgroups are not equal, each control limit is not a single straight line but varies with the subgroup size.

Click on the P Chart Options button and the Parameters tab if you wish to enter a historical value for p.

P Chart - Options

Historical Groups | Parameters | Estimate | Tests | S Limits | Storage

Historical proportion:

Constant(s): .12

Select

Help | OK | Cancel

The p chart shows a clear downward trend in the daily proportion of workers who are absent. It appears that actions were taken to reduce the absenteeism rate. The last two weeks' data suggest that the rate has stabilized.

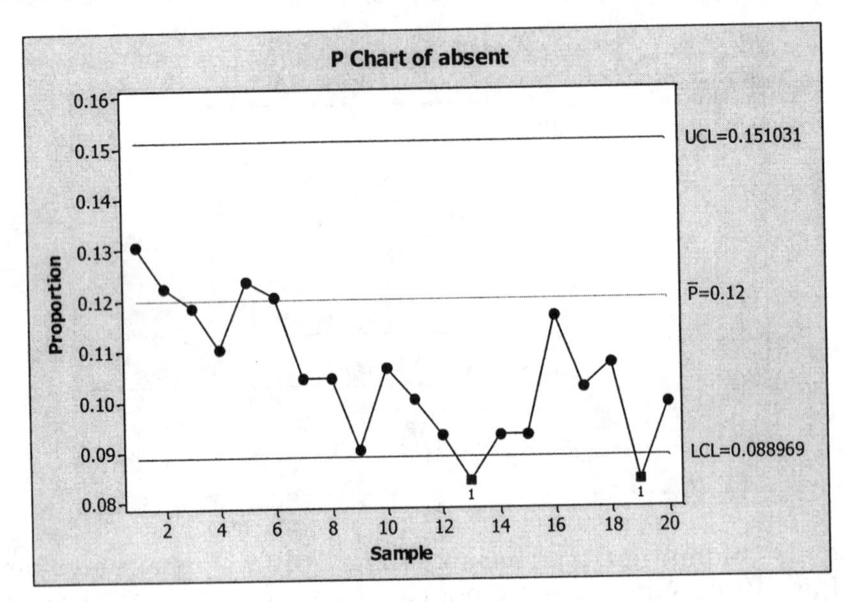

EXERCISES

24.10 A pharmaceutical manufacturer forms tablets by compressing a granular material that contains the active ingredient and various fillers. The hardness of a sample from each lot of tablets is measured in order to control the compression process. The process has been operating in control with mean at the target value μ = 11.5 and estimated standard deviation σ = 0.2. Table 24.2 gives three sets of data, each representing $\bar{x}$ for 20 successive samples of n = 4 tablets. One set remains in control at the target value. In a second set, the process mean μ shifts suddenly to a new value. In a third, the process mean drifts gradually.

(a) What are the center line and control limits for an $\bar{x}$ chart for this process?

(b) Select **Stat ➤ Control Charts ➤ Variables for Subgroups ➤ Xbar** to draw separate $\bar{x}$ charts for each of the three data sets. Mark any points that are beyond the control limits.

(c) Based on your work in (b) and the appearance of the control charts, which set of data comes from a process that is in control? In which case does the process mean shift suddenly and at about which sample do you think the mean changed? Finally, in which case does the mean drift gradually?

24.13 Exercise 24.10 concerns process control data on the hardness of tablets for a pharmaceutical product. Table 24.4 gives data for 20 new samples of size 4, with the $\bar{x}$ and s for each sample. The process has been in control with mean at the target value $\mu = 11.5$ and standard deviation $\sigma = 0.2$.

(a) Select **Stat ➤ Control Charts ➤ Variables for Subgroups ➤ Xbar-S** to make both $\bar{x}$ and s charts for these data based on the information given about the process.

(b) At some point, the within-sample process variation increased from $\sigma = 0.2$ to $\sigma = 0.4$. About where in the 20 samples did this happen? What is the effect on the s chart? On the $\bar{x}$ chart?

(c) At that same point, the process mean changed from $\mu = 11.5$ to $\mu = 11.7$. What is the effect of this change on the s chart? On the $\bar{x}$ chart?

24.15 Figure 24.10 reproduces a data sheet from the floor of a factory that makes electrical meters. The sheet shows measurements of the distance between two mounting holes for 18 samples of size 5. The heading informs us that the measurements are in multiples of 0.0001 inch above 0.6000 inch. That is, the first measurement, 44, stands for 0.6044 inch. All the measurements end in 4. Although we don't know why this is true, it is clear that in effect the measurements were made to the nearest 0.001 inch, not to the nearest 0.0001 inch. EX24-15.MTW contains the data along with $\bar{x}$ and s for all 18 samples. Based on long experience with this process, you are keeping control charts based on $\mu = 43$ and $\sigma = 12.74$. Select **Stat ➤ Control Charts ➤ Variables for Subgroups ➤ Xbar-S** to make s and $\bar{x}$ charts for the data in Figure 24.10 and describe the state of the process.

24.21 Table 24.6 in BPS and TA24-06.MTW give data on the losses (in dollars) incurred by a hospital in treating DRG 209 (major joint replacement) patients. The hospital has taken from its records a random sample of 8 such patients each month for 15 months. Select **Stat ➤ Control Charts ➤ Combination Charts ➤ Xbar-S** from the menu to make an $\bar{x}$ and s chart.

(a) Does the s control chart show any points out of control? Is it save to base the $\overline{x}$ chart on all 15 samples?

(b) Is the $\overline{x}$ chart in control?

24.25 Table 24.6 of BPS and EX2406.MTW give data on hospital losses for samples of DRG 209 patients. Select **Stat ➤ Basic Statistics ➤ Display Descriptive Statistics** from the menu to obtain numerical summaries for the data. The distribution of losses has been stable over time. What are the natural tolerances within which you expect losses on nearly all such patients to fall?

24.26 Do the losses on the 120 individual patients in Table 24.6 in BPS and TA24-06.MTW appear to come from a single Normal distribution? Select **Graph ➤ Histogram** from the menu and select "With Fit" and discuss what the graph shows. Are the natural tolerances you found in the previous exercise trustworthy?

24.27 If the mesh tension of individual monitors follows a Normal distribution, we can describe capability by giving the percent of monitors that meet specifications. The old specifications for mesh tension are 100 to 400 mV. The new specifications are 150 to 350 mV. Because the process is in control, we can estimate that tension has mean 275 mV and standard deviation 38.4 mV. Select **Calc ➤ Probability Distributions ➤ Normal** to answer the following questions.

(a) What percent of monitors meet the old specifications?

(b) What percent meet the new specifications?

(c) We can improve capability by adjusting the process to have center 250 mV. This is an easy adjustment that does not change the process variation. What percent of monitors now meet the new specifications?

24.31 After inspecting Figure 24.16 in BPS, you decide to monitor the next four weeks' absenteeism rates using a center line and control limits calculated from the second two weeks' data recorded in Table 24.8 and TA24-08.MTW. Find $\overline{p}$ for these 10 days. Select **Stat ➤ Control Charts ➤ Attributes chart ➤ P** from the menu. Click on the P Chart Options button and the Parameters tab to use this value to compute the CL, LCL, and UCL. (Until you have more data, these are trial control limits. As long as you are taking steps to improve absenteeism, you have not reached the process-monitoring stage.)

24.35 Here and in EX24-35.MTW are data from an urban school district on the number of eighth-grade students with three or more unexcused absences from school during each month of a school year. Because the total number of eighth graders changes a bit from month to month, these totals are also given for each month.

Month	Sept.	Oct.	Nov.	Dec.	Jan.	Feb.	Mar.	Apr.	May	June
Students	911	947	939	942	918	920	931	925	902	883
Absent	291	349	364	335	301	322	344	324	303	344

(a) Select **Calc ➤ Calculator** from the menu to find $\bar{p}$. Because the number of students varies from month to month, also find $\bar{n}$, the average per month.

(b) Select **Stat ➤ Control Charts ➤ Attributes chart ➤ P** to make a p chart using control limits based on $\bar{n}$ (rounded) students each month. Comment on control.

(c) The exact control limits are different each month because the number of students n is different each month. This situation is common in using p charts. Instead of using $\bar{n}$ for the Subgroup "Size", you can select a Subgroup "Indicator Column" when you make your chart. Does using exact limits affect your conclusions?

24.39 You manage the customer service operation for a maker of electronic equipment sold to business customers. Traditionally, the most common complaint is that equipment does not operate properly when installed, but attention to manufacturing and installation quality will reduce these complaints. You hire an outside firm to conduct a sample survey of your customers. Here and in EX24-39.MTW are the percents of customers with each of several kinds of complaints:

Category	Percent
Accuracy of invoices	25
Clarity of operating manual	8
Complete invoice	24
Complete shipment	16
Correct equipment shipped	15
Ease of obtaining invoice adjustments/credits	33
Equipment operates when installed	6
Meeting promised delivery date	11
Sales rep returns calls	4
Technical competence of sales rep	12

(a) Why do the percents not add to 100%?

(b) Select **Graph ➤ Bar Chart** to make a Pareto chart. What area would you choose as a target for improvement?

24.43 Painting new auto bodies is a multistep process. There is an "electrocoat" that resists corrosion, a primer, a color coat, and a gloss coat. A quality study for one paint shop produced this breakdown of the primary problem types for those autos whose paint did not meet the manufacturer's standards:

Problem	Percent
Electrocoat uneven—redone	4
Poor adherence of color to primer	5
Lack of clarity in color	2
"Orange peel" texture in color coat	32
"Orange peel" texture in gloss coat	1
Ripples in color coat	28
Ripples in gloss coat	4
Uneven color thickness	19
Uneven gloss thickness	5
Total	100

Select **Graph ➤ Bar Chart** to make a Pareto chart. Which stage of the painting process should we look at first?

24.48 Table 24.9 in BPS and TA24-09.MTW give process control samples for a study of response times to customer calls arriving at a corporate call center. A sample of 6 calls is recorded each shift for quality improvement purposes. The time from the first ring until a representative answers the call is recorded. Table 24.9 gives data for 50 shifts, 300 calls total. Table 24.9 also gives $\bar{x}$ and s for each of the 50 samples.

(a) Select **Stat ➤ Control Charts ➤ Variables Charts ➤ Xbar-S** from the menu to make an $\bar{x}$ and s chart. Check for points out of control on the s chart.

(b) If the s-type cause responsible is found and removed, then we can remove the points that were out of control on the s chart. Select **Stat ➤ Control Charts ➤ Variables Charts ➤ Xbar-S** from the menu to make new charts. Find the new control limits and verify that no points s are now out of control.

(c) Comment on the control (or lack of control) of $\bar{x}$ of the remaining 46 samples. (Because the distribution of response times is strongly skewed, s is large and the control limits for $\bar{x}$ are wide. Control charts based on Normal distributions often work poorly when measurements are strongly skewed.)

Appendix
Minitab Session Commands

Topics to be covered in this chapter:

Session Commands and the Session Window
Rules for Entering Session Commands
Command Prompts
Session Command Syntax and Menu Equivalents

Session Commands and the Session Window

Most functions in Minitab are accessible through menus, as well as through a command language called session commands. You can use menu commands and session commands interchangeably, or you can use one of the two exclusively. Menu commands provide clickable options through menus and dialog boxes. Session commands allow you to provide specific instructions, through a command language. Most session commands are simple, easy to remember words like PLOT, SAVE, or SORT.

The Session window is primarily used for displaying the results of commands, as text. However, you can also type session commands in the Session window shown here by turning on the MTB> command prompt.

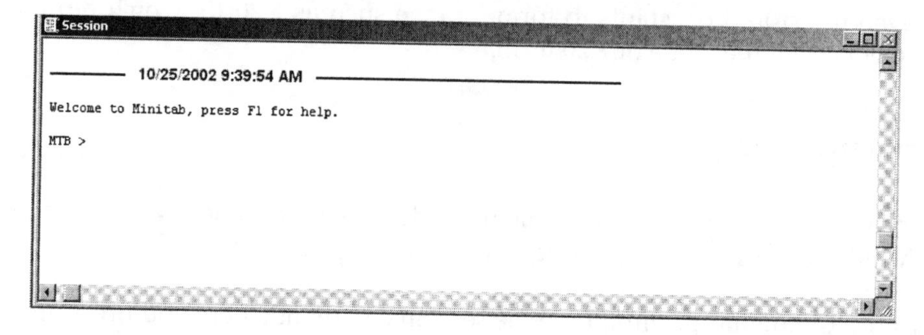

To turn on the MTB> command prompt, click on the Session window and select

Editor ➤ Enable Command Language

from the Minitab menu. If you pull down the Editor menu, there is a check box to the left of Enable Command Language. If there is already a check, selecting

Editor ➤ Enable Command Language will disable the command language. Type commands at the MTB> prompt in the last line of the Session window.

Rules for Entering Session Commands

A session command consists of one main command and may have one or more subcommands. Arguments and symbols may also be included in the command. Subcommands, which further define how the main command should be carried out, are usually optional. Arguments specify data characteristics.

To execute a command, type the main command followed by any arguments. If the command has subcommands, end the command line with a semicolon. Type subcommands at the SUBC> prompt. Put a semicolon (;) after each subcommand. Put a period (.) after the last subcommand. Press <Enter> to execute a command.

Commands and column names are not case-sensitive; you can type them in lowercase, uppercase, or any combination. You can abbreviate any session command or subcommand by using the first four letters.

Arguments specify data characteristics, such as location or titles. They can be variables (columns or constants) as well as text strings or numbers. Enclose variable names in single quotation marks (for example, HISTOGRAM 'Salary'). In arguments, variable names and variable numbers can be used interchangeably. For example, DESCRIBE C1 C2 and DESCRIBE 'Sales' C2 do the same thing if C1 is named 'Sales.'

You can abbreviate a consecutive range of columns, stored constants, or matrices with a dash. For example, PRINT C2-C5 is equivalent to PRINT C2 C3 C4 C5. You can use a stored constant (such as K20) in place of any constant. You can even use stored constants to form a range such as K20:15, which represents all integers from the value of K20 to 15.

Command Prompts

The prompts that appear in the Session window help you know what kind of input Minitab expects. There are five different prompts:

MTB> Command prompt; type the session commands here and press Enter.
SUBC> Subcommand prompt; type the subcommands here or type ABORT to cancel the entire command.
DATA> Data prompt; enter data here. To finish entering data and return to the MTB> prompt, type END and press Enter.
CONT> Continuation prompt; if the command from your previous line ends with the continuation symbol &, Minitab displays CONT> on the next line so you can enter the rest of the command or data.

Session Command Syntax and Menu Equivalents

In the following, commands are listed by function. In the session command syntax, K denotes a constant such as 8.3 or k14, C denotes a column, such as C12 or 'Height,' and E denotes either a constant or column. Square brackets [ ] enclose optional arguments. Menu equivalents follow each command or group of commands.

General Information

HELP `command`
Help ➤ Search for Help on

INFO `[C...C]`
menu equivalent not available

STOP
File ➤ Exit

Managing Data

SET `data into C`
Calc ➤ Make Patterned Data

INSERT `data [between rows K and K] of C...C`
Editor➤ Insert Cells

END `of data`
menu equivalent not available

NAME `E = 'name' ... E = 'name'`
In the Data window, click a column name cell and type the name

PRINT `the data in E...E`
Data ➤ Display Data

SAVE `[in file in "filename" or K]`
File ➤ Save Worksheet (As)

RETRIEVE `[file in "filename" or K]`
File ➤ Open Worksheet

Editing and Manipulating Data

CODE `(K...K) to K ... (K...K) to K for C...C, put in C...C`
Data ➤ Code

DELETE `rows K...K of C...C`
Data ➤ Delete Rows

ERASE E...E
Data ➤ Erase Variables

INSERT data [between rows K and K] of C...C
Editor➤ Insert Cells

LET C(K) = K
Calc ➤ Calculator

SORT C [carry along C...C] put into C [and C...C]
Data ➤ Sort

STACK (E...E) on ... on (E...E), put in (C...C)
Data ➤ Stack ➤ Columns

UNSTACK (C...C) into (E...E) ... (E...E)
Data ➤ Unstack Columns

Arithmetic

LET E = expression
ADD E to E...E, put into E
SUBTRACT E from E, put into E
MULTIPLY E by E...E, put into E
DIVIDE E by E, put into E
RAISE E to the power E put into E
ABSOLUTE value of E put into E
SQRT of E put into E
LOGE of E put into E
LOGTEN of E put into E
EXPONENTIATE E put into E
ANTILOG of E put into E
ROUND E put into E
Calc ➤ Calculator

CENTER the data in C...C put into C...C
Calc ➤ Standardize

COUNT the number of values in C [put into K]
N count the nonmissing values in C [put into K]
NMISS (number of missing values in) C [put into K]
SUM of the values in C [put into K]
MEAN of the values in C [put into K]
STDEV of the values in C [put into K]
MEDIAN of the values in C [put into K]
MINIMUM of the values in C [put into K]
MAXIMUM of the values in C [put into K]
Calc ➤ Column Statistics

RCOUNT of E...E put into C
RN of E...E put into C
RNMISS of E...E put into C
RSUM of E...E put into C

RMEAN of E...E put into C
RSTDEV of E...E put into C
RMEDIAN of E...E put into C
RMINIMUM of E...E put into C
RMAXMUM of E...E put into C
Calc ➤ Row Statistics

Distributions and Random Data

RANDOM K observations into C...C
Calc ➤ Random Data

PDF for values in E...E [put results in E...E]
Calc➤ Probability Distributions

CDF for values in E...E [put results in E...E]
Calc ➤ Probability Distributions

INVCDF for values in E [put into E]
Calc ➤ Probability Distributions

SAMPLE K rows from C...C put into C...C
Calc ➤ Random Data ➤ Sample From Columns

Graphics

BOXPLOT of C...C
Graph ➤ Boxplot

CHART C...C
Graph ➤ Chart

HISTOGRAM of C...C
Graph ➤ Histogram

STEM-AND-LEAF of C...C
Graph ➤ Stem-and-Leaf

PIECHART C...C
Graph ➤ Pie Chart

PLOT C vs C
Graph ➤ Scatterplot

TSPLOT [period = K] of C
Graph ➤ Time Series Plot

Basic Statistics

`CORRELATION C...C`
Stat ➤ Basic Statistics ➤ Correlation

`DESCRIBE variables in C...C`
Stat ➤ Basic Statistics ➤ Descriptive Statistics

`ONET C...C`
Stat ➤ Basic Statistics ➤ 1-Sample t

`ONEZ C...C`
Stat ➤ Basic Statistics ➤ 1-Sample Z

`PAIR C C`
Stat ➤ Basic Statistics ➤ Paired t

`PONE C...C or K K...K`
Stat ➤ Basic Statistics ➤ 1 Proportion

`POWER`
Stat ➤ Power and Sample Size

`PTWO C C or K K K K`
Stat ➤ Basic Statistics ➤ 2 Proportions

`TWOSAMPLE test and CI [K% confidence] samples in C C`
Stat ➤ Basic Statistics ➤ 2-Sample t

`TWOT test with [K% confidence] data in C, groups in C`
Stat ➤ Basic Statistics ➤ 2-Sample t

`CORRELATION between C...C`
Stat ➤ Basic Statistics ➤ Correlation

Regression

`REGRESS C on K predictors C...C`
Stat ➤ Regression ➤ Regression

`FITLINE y in C, predictor in C`
Stat ➤ Regression ➤ Fitted Line Plot

Analysis of Variance

`ANOVONEWAY for samples in C...C`
Stat ➤ ANOVA ➤ Oneway (unstacked)

`ONEWAY data in C, levels in C`
Stat ➤ ANOVA ➤ Oneway

Nonparametrics

KRUSKAL-WALLIS `test for data in C, levels in C`
Stat ➤ Nonparametrics ➤ Kruskal-Wallis

MANN-WHITNEY `two-sample rank test with [K% confidence] on C C`
Stat ➤ Nonparametrics ➤ Mann-Whitney

WTEST `one-sample rank test [of median = K] on C...C`
Stat ➤ Nonparametrics ➤ 1-Sample Wilcoxon

Tables

TALLY `the data in C...C`
Stat ➤ Tables ➤ Tally

TABLE `the data classified C...C`
Stat ➤ Tables ➤ Cross Tabulation and Chi-Square

CHISQUARE `test on table stored in C...C`
Stat ➤ Tables ➤ Chi-Square Test (Table in Worksheet)

Quality Control

ICHART `for C`
Stat ➤ Control Charts ➤ Variables Chart for Individuals ➤ Individuals

PCHART `number of nonconformities are in C...C, sample size = E`
Stat ➤ Control Charts ➤ Attributes chart ➤ P

SCHART
Stat ➤ Control Charts ➤ Variables Chart for Subgroups ➤ S

XBARCHART
Stat ➤ Control Charts ➤ Variables Chart for Subgroups ➤ Xbar

XSCHART
Stat ➤ Control Charts ➤ Variables Chart for Subgroups ➤ Xbar-S

Index